E. Bud Guitrau

The EDM Handbook

HANSER

Hanser Publishers, Munich   Hanser Publications, Cincinnati

The Author:
E. Bud Guitrau, 1484 E. Garcia Place, PLACENTIA, CA 92870, USA

Distributed in the USA and in Canada by
Hanser Publications
6915 Valley Avenue, Cincinnati, Ohio 45244-3029, USA
Fax: (513) 527-8801
Phone: (513) 527-8896 or 1-800-950-8977
www.hanserpublications.com

Distributed in all other countries by
Carl Hanser Verlag
Postfach 86 04 20, 81631 München, Germany
Fax: +49 (89) 98 48 09
www.hanser.de

Guitrau, E. P.
    The EDM handbook / by E. P. Guitrau.
    312 p. 17.75 × 25.4 cm.
    Includes bibliographical references and index.
    ISBN 1-56990-242-9
    1. Electric metal-cutting--Handbooks, manuals, etc. I. Title.
TJ1191.G75 1997
671.5'3--dc21   97-35552
                                                    CIP

Bibliografische Information Der Deutschen Bibliothek
Die Deutsche Bibliothek verzeichnet diese Publikation in der Deutschen Nationalbibliografie; detaillierte bibliografische Daten sind im Internet über <http://dnb.d-nb.de> abrufbar.

ISBN 978-3-446-42046-5

© Carl Hanser Verlag, Munich 2009
Production Management: Steffen Jörg
Coverconcept: Marc Müller-Bremer, Rebranding, München, Germany
Coverdesign: Stephan Rönigk
Printed and bound by Kösel, Krugzell
Printed in Germany

CONTENTS

PREFACE

This book will not tell you how to operate your machine. That is a wholly separate book in itself, best left to those who do that sort of thing. This book attempts to give the reader a better understanding of what EDM is and how it works. Only after knowing and understanding this can you make EDM work for you. Only if you understand its theory can you practice it, and only then can you know the premise within which it operates and, in turn, understand the promise it can deliver.

EDM. Eat Dat Metal, Evil Darn Machine, Extra Dry Martini: regardless of how you feel about it, whatever your opinion, this is the book you should read if you want to know more about what is involved in the world of EDM.

This book was written for every student and apprentice, for I had many books while studying but, unfortunately, none of this craft.

This book was written for every operator, machinist, toolmaker, and journeyman. The technology is not daunting. You are already professionally engaged in one of the most demanding crafts. Why would you shun learning additional specialized facets of the craft, and increase your own personal knowledge and experience along the way?

This book was written for every engineer, designer, and manufacturing engineer who is reluctant to take the bold leap of faith and accept EDM as one of the fastest, most accurate, most economical methods of manufacture (regardless of what you may have heard). You will be surprised to find that, many times, it is.

This book was written to help all readers understand EDM. To understand its theory and practice. To understand its capabilities as well as its limitations. To understand that for every nugget revealed, there is often a landmine hidden somewhere else. This is a crash-course on EDM, hopefully educating the observer and aiding the operator. This book works if the reader becomes more capable of making or evaluating an EDM solution.

On some technical issues I have taken some liberties in striving to make them more understandable, although they may not be wholly or technically correct. Most of these will go unnoticed and none will negatively affect your actions or results. Some professors and theorists may notice some of these and wish to take exception, but I would rather they understand my intent and the context within it was written.

I would like to thank everyone who has helped and supported me along the way. I could never hope to name everyone but you already know who you are. My family in particular deserves a special thank you. They have endured my long evenings spent reading, researching and writing. I already travel quite a bit, which is difficult enough on wife and children, but they have taken my additional pursuits in stride and have always supported me despite the sometimes unfairness of the situation.

I must also thank you, the reader, for your interest and support. Please continue your quest for EDM knowledge and information. We must *"Learn to burn."*

ABOUT THE AUTHOR

E. Bud Guitrau works for a major EDM company and has been involved in EDM for almost 25 years. As a writer, he was on the staff of EDM TODAY magazine for four years and most recently helped complete the EDM section of the latest edition of *Machinery's Handbook*.

SECTION I

INTRODUCTION TO EDM

Chapter 1

AN OVERVIEW OF TECHNOLOGY AND TRENDS

INTRODUCTION

Since the introduction of EDM over fifty years ago, the technology for speed and precision has grown tremendously and, likewise, its capabilities and manufacturing applications have followed suit. Expanding from its initially small niche of precision tooling manufacture, its broad capabilities have allowed it to expand its influence to encompass production, aerospace/aircraft, medical, and virtually all areas of conductive material machining. Further augmented by automatic tool changers, automatic threading, slug removers, robotic workpiece changers, and palletization, this fascinating manufacturing process has attained almost virtual machining autonomy and has rightfully taken its place alongside the more conventional machining processes of lathes, mills, and grinders as self-supporting profit centers.

VERTICAL MACHINES

Originally known as "vertical," "ram," or "sinker" (so named after their earlier, primary toolroom operation of die-sinking) EDM because the electrode was mounted and used in the vertical or Z-axis only, the first working models were adapted to drill presses and small milling machines. Their primary duties were "busting taps" — removing broken taps and drills from expensive workpieces. In time, the power supplies grew in sophistication, and warranted being attached to their own "stand-alone" machine tool.

3

At first, EDM was only used in an "emergency," and then mostly in toolroom applications. Today's modern EDM equipment is a far cry from its primitive beginnings. Very precise hydraulic servos became necessary and were developed as "adaptive controls" became more prevalent. As EDM's came into greater use, they were soon outfitted with NC controllers and then CNC. Today's CNC EDM's can be equipped with "fuzzy logic" controls, tool and workpiece changers, and can drive up to six axes simultaneously.

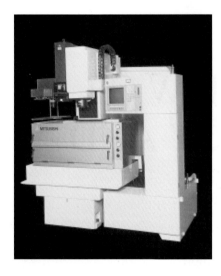

Figure 1.1   A modern CNC EDM equipped with automatic tool changer and C-axis.

THE ELECTRIC BAND SAW

Although this description paints a very poor picture of the degree of accuracy and finishes attainable by wire EDM, this comparison does enable a layman to envision the concept of a wire-cut EDM. Much like a band saw in principle, the wire (saw blade) is fed from a spool and passes completely through the workpiece. The wire is electrically charged and "cuts" through the workpiece by "spark erosion," vaporizing and melting the material instead of cutting mechanically like the band saw. Since the "cutter" (wire) never physically touches the workpiece, there are no cutting pressures or mechanical stresses produced to influence the part or the setup. Further, this process provides burr-free multiaxis machining of many parts that would not otherwise be possible due to material hardness or shape. This process however, is much more refined and elegant than the reference to an "electric band saw."

POWER SUPPLIES

The continuing evolution of the power supply (or generator) is perhaps the single greatest reason for the rapid advancement in cutting speeds, especially in wire EDM. It is interesting to look at the different methods and approaches taken by the builders in this ever-increasing quest for speed. Ironically, improvements in the wire electrode itself, such as stratified and coated wires, allowed the full poten-

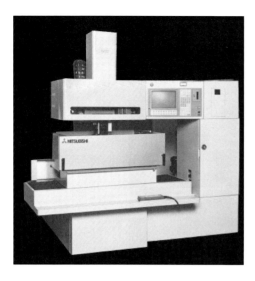

Figure 1.2   This wire EDM is equipped with automatic wire threader and submerged cutting.

tial of the existing power supplies to be used. This came about a few years ago when some power supplies were running at or, in some cases, exceeding their rated peak currents and, as a result, they were burning transistor boards and switching units. This forced the wire EDM engineers and builders to improve their power supplies to live up to the challenge that the new wires were presenting. Improvements beget improvements.

Vertical or sinker technology continued to improve and, like the wires, were helped substantially by improved electrode materials, particularly graphite. The development of high-density isotropic graphites replaced copper electrodes, reducing electrode wear considerably while becoming easier to fabricate, thereby reducing EDM operating costs.

With sinkers, the amount of power or amperage is relative to the size of the machine, and this holds true for most builders. Some wire machine builders provide power supplies with up to 40 amps capability. Another builder uses less than 20 amps in conjunction with modified spark wave-forms to accomplish the same amount of work. All are successful.

Some of the modern high-performance wire-cut machines are successfully cutting over 30 square inches per hour in tool steels. Cutting speeds in aluminum have exceeded 80 square inches! Who would have imagined that cutting speeds of wire EDM would be over *30 times faster* than they were just 15 years ago?

These significant increases in cutting speeds have allowed wire-cutting to encroach upon some of the more traditional milling operations which, simply by their nature, can *create* second and third operations to complete. (See Summary Analysis of Airbus Actuator in Appendix A.)

High-speed cutting is not the only quality required by wire EDM. There are many variables to contend with, such as entry cuts and sharp corners. These require adaptive controls with the ability to "see ahead" many blocks of program code to adapt to approaching sharp corners, calculate mathematical program variables, or to reduce the current and/or the frequency of the spark to allow accurate machining without breaking the wire. The faster the cutting speeds and greater the accuracies, the faster the computer needs to be able to identify a situation and effect change to stabilize the machining.

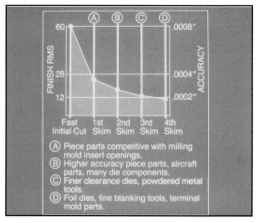

Figure 1.3   Wire EDM's ability to compete with conventional machining is shown by this chart.

CONTROL SYSTEMS

There are almost as many drive systems as there are makes of machines. Some builders use stepper motors and others use servo motors, both ac and dc. At present, encoder resolutions on some machines are as fine as 0.1 microns (or 0.000004"), with this movement being verified by closed-loop feed back or glass scales. Almost all of the newest machines are capable of moving in inch or metric modes with digital displays of five and four decimal places, respectively. These combinations are being used to drive up to five and six axes, simultaneously.

On vertical machines so equipped, these would be:
the traditional X- and Y-axes (the worktable),
the Z-axis,
an indexable, servo-controlled C-axis (spindle),
an external, servo-controlled "A"- or "B"-axis.

Wire machines can be configured as follows:
the traditional X- and Y-axes (the worktable),
the upper and lower control arms (the U- and V-axes),
the "traveling Z-axis,"
an external, servo-controlled "A"- or "B"-axis.

Both types of machines can routinely machine parts within one or two "tenths" (0.0001"), and higher-accuracy models can machine to "split-tenth" accuracy.

To support movement in increments this fine, the computer's processing speed was increased to deliver the full potential these drivers can provide. Over the past few years, the trend in controls has been the progression from 16-bit to 32-bit processors for increased computing speed. To the layman, the necessity for 32-bit processing may not seem clear, but a simple analogy would be, "Which would move more traffic, a highway with 16 or 32 lanes?" This allows 32 bits of coded information to be processed simultaneously for faster response to the demanding conditions faced by adaptive controls, and screen refresh rates of complex computer graphics. Clock speeds are up to 24.5 MHz on some controls; so now, not only is the "highway" much "wider," but the "speed-limit" has been increased as well.

This combination of processing speed and power allows the computer to precisely monitor and control servo voltages and feed rates, attenuate and dampen wire harmonics, and to provide immediate response to constant changes in cutting conditions through high-speed adaptive controls. The good just keeps getting better.

MACHINE TOOLS

The machine tool continues its evolution as builders learn from experience and strive to get closer and closer to perfection. Computer-designed structures of high-quality seasoned castings make up the backbone of the machine, with strategic areas being heavily ribbed and gusseted for strength and stability. One builder departs from traditional design, and instead of using iron castings, constructs their machine frame out of an inert, cement-like material. Both wire and sinker use various frame designs: C-frames and bridge or gantry styles. Both types have been fitted with ceramic components for rigidity and low coefficient of expansion.

Whether movement is done by the worktable or the gantry, most CNC machines are controlled through precision-ground recirculating ball screws, totally eliminating errors due to backlash. Movements are usually over glass-smooth linear-motion slideways for minimal inertia resistance. This is very important to help achieve the fine finishes and high accuracies demanded today.

Vertical machine tool sizes range from small, bench-top models to huge behemoths capable of EDMing workpieces the size of automobile fender and hood dies and dashboard molds. Wire machines likewise run from small table-top units to units capable of accommodating part sizes of 40" by 64" and 18" thick. Special machines and modifications have successfully wire-cut parts over 50" tall. Consider the degree of technology and sensitivity required to monitor and control a moving workpiece weighing over two tons as it is being precisely cut by a brass wire only 0.012" in diameter!

Some wire machines cut only in two axes — the X and Y. These are for two-dimensional or straight-walled parts. Others are standard with four axes — the traditional X- and Y-axes with the addition of a U- and V-axes, the upper and lower control arms, respectively. This allows the machines to cut tapered shapes, for example, the angular reliefs required on die openings and cutting tools.

Most of us have seen the unique wire EDMed sample parts displayed at machine tool shows and sales demonstrations. These typically show the machine's ability to cut two separate programs — one on the top surface of the workpiece and one on the bottom. These parts, although seen in a myriad shapes, are often referred to as "squircles" because a square can be executed on top and a circle executed on the bottom.

Tapering capabilities usually average from 10 to 15 degrees, which will encompass, by estimate, over 95% of most taper requirements, but some machines specialize in steep tapering capabilities. This is usually done with the U- and V-axes incorporated within articulated heads that can swivel to angles of 30 degrees or more, keeping the wire guides tangent to the travel of the wire.

Many machines can drive fifth and sixth axes. The fifth axis is a programmable Z-axis, giving the machine the ability to raise and lower the upper guides across an uneven workpiece. This keeps the guides and nozzles as close to the workpiece as possible for better accuracy and flushing. The sixth axis can be an auxiliary driver such as a rotary table, tilt table, indexer, or other such device. This offers the creative and imaginative designer and engineer almost unlimited capability in doing that "impossible" part.

All manual vertical EDM's use the three primary axes: X, Y, and Z. CNC units drive these three and others — five or six as standard, and, on some models, even more. Many CNC EDM machines come equipped with a spindle-oriented C-axis. This feature allows the unit to perform indexing and helical machining as in machine-tapping of threads or cutting helical gears in prehardened materials. The fifth and sixth axes can be remote units, placed on the worktable and used as tooling as with servo controlled tilt tables, rotary tables, and B-axis devices.

HIGH-PRESSURE FLUSHING
AND SUBMERGED CUTTING

The old saying, "The key to successful EDMing is *flushing, flushing, and flushing,*" is not an adage belonging only to the vertical EDM. While in most instances flushing is an easier subject to address in wire-cut operations, nevertheless, it is still imperative for efficient operations.

Deionized water is used in wire-cut EDM and is an integral part of EDM theory. The dielectric fluid provides insulation against premature discharging, cools the machined area, and flushes away the "chips." Without flushing, EDM simply will not work. Special nozzles are used to direct a stream or jet of water around the wire and through the workpiece.

We have already given partial credit for the fast cutting speeds of today to the engineers and builders who continually "soup-up" the power supplies, but the fastest power supply in the world wouldn't be so without the advent of high-pressure flushing. Today's flushing pressures are running at 300 psi and are absolutely necessary to flush the large amounts of swarf from the narrow kerf left by the wire.

Ultimately, as the flushing pressures kept rising higher and higher on the newer machines, certain jobs that had previously been successful on older machines of the same make had difficulty maintaining finishes and accuracies on the newer models. It was eventually determined that certain fixtures, tooling, and workholding devices would encounter harmonic vibrations or actually deflect under the high flushing pressures. This prompted commercial tooling companies to "beef-up" and strengthen existing and pending designs. Once again, improvements in one area of the process warranted improvements in another to realize the total benefit. In this way, the entire EDM industry benefits from faster, more accurate, and more reliable products.

High-pressure flushing has been the solution for high-speed cutting, but it did not solve all of the flushing-related problems. Steep tapers, interrupted cuts, aircraft "honeycomb," and round or tubular parts are just a few examples of applications that pose flushing problems. In these instances, flushing is inconsistent and perhaps nonexistent, so speeds decrease and wire breaks are prevalent. Under these conditions, it is recommended that the part be submerged during machining. Submerged cutting greatly improves the poor conditions encountered in these and other examples for faster, more reliable machining.

Submerged wire-cutting has been used almost as long as wire machines have been made, but initially only by a few innovative jobbers, and eventually one or two machine builders. Today, submerged cutting is a common procedure with most manufacturers offering submerged models. After-market "sub-tanks" are also available for users with older, nonsubmergable machines or limited budgets.

Submerged wire-cutting offers another advantage over nonsubmerged — changes in room temperatures will have less influence on accuracy since the entire workpiece, control arms, worktable, and tooling are all submerged in the same temperature-controlled dielectric.

UNATTENDED AND AUTOMATED OPERATIONS

One of the most appealing features that EDM presents to manufacturing is unattended machining. The ability for a job to run far into the night or over a weekend goes a long way in helping a plant or shop justify its purchase.

The most common item used to support unattended and overnight operations is the automatic wire threader. An auto-threader is a device that will automatically rethread the wire in the event of an accidental wire break or in the case of a workpiece with multiple openings, like those often found in a progressive die.

For example: multiple programs can be written and linked, and the wire machine will move to the first opening, thread the wire, cut the first detail, sever its own wire, move to the next opening, rethread, and continue machining far into the night.

Auto-threaders are becoming increasingly popular but they haven't always been well received. When first introduced, they were a source of frustration to machine operators due to poor reliability, a high degree of maintenance, and most of them were just plain slow. Automatic wire threaders are typically very complex and precise mechanisms. Over the years, many refinements have provided us with today's threaders that are fast, reliable, and give good overall performance. They are most useful for jobs with multiple detail openings such as progressive dies, or in production operations where multiple parts are fixtured and movement and rethreading is required from part to part.

Automatic wire threaders are the first step toward total automation of wire EDM, and statistics indicate that in 1993 over 60% of new machine orders specified autothreaders (up from approximately 40% in 1991).

There are several types of threading devices: water jet, pipe, and combinations of both. These feed the wire through the lower guide to a retrieval system of drive belts or vacuum assisted pinch rollers. The spent wire is then fed into a waste bin or take-up spool. In the event of a misfeed or blocked wire-start hole, machines can be programmed to retry the start-hole for a specific number of times. If the threader is unable to successfully thread after the programmed number of attempts have been completed, the machine will move on to the next position, leaving *only that detail* unfinished instead of the rest of the part.

One of the problems presented by wire-cutting is the removal of the central core or slug that is left when cutting the part (for example, a die opening). If even a small slug is allowed to fall freely, the least it can do is pinch and break the wire. Worst-case scenarios can have it become jammed in the lower nozzle, resulting in broken nozzles or damage to the part, tooling, or the lower control arm. Today's higher cutting speeds have allowed "pocketing" routines to be used, effectively eliminating the slug altogether, but this is used only on small details that do not consume a lot of time.

On the larger details, optional slug removers can be used. When the part has been completely roughed and skimmed (except for strategically sized and located cutoff tabs), the wire is automatically cut and the lower arm traverses away from this area. A piston or punch then knocks the slug from the opening from above. On some machines, this slug drops down below the workpiece onto a conveyer system which removes it from the machining envelope. Other models have a mechanical arm to "rake" the slug into a waste bin. Another uses the lower arm itself to push the slug out of the work area.

Some builders remove wire-cut slugs from workpieces by lifting them out. In one method, a hole is drilled in the slug-to-be, prior to wire-cutting. The part is machined leaving the cutoff tabs, as with the other styles, but the slug is removed by an expanding mandrel that is fed into the hole and enlarged to grip the inside of the hole. An actuator located above the workpiece then lifts the slug up and out of the cavity for disposal and to allow finish machining of the cutoff tabs and any remaining skim cuts. Other "lifting" models are similar, except a magnetic arm is used to lift out the slug. The units using a magnetic retrieval system are obviously limited to lifting only magnetic materials.

If automatic slug removal is to be used successfully, it must be examined beforehand and designed into the manufacturing process. First, the design of the part or workpiece must lend itself to automatic slug removal. The part's shape must allow for strategically placed cutoff tabs to "balance" the slug and prevent it from tipping or "racking," i.e., wedging it in the workpiece. Obviously, if the slug is too large or heavy, it will have to be removed manually.

In pursuit of greater automation, pallets and pallet changers have been in use for some time. Palletization is usually seen in larger companies with machines dedicated to production operations on long-running jobs. There are many types of pallet changers and conveyer systems used in FMS and work cells where the pallets move from set-up stations, to machining centers, EDM's, and finally on to CMM's for totally unattended manufacturing from the blank workpiece to its final inspection.

Robotic equipped wire EDM is more prevalent in Europe and Japan than in the United States. This can be attributed, in part, to the larger numbers of CNC units sold there, although an even higher number of robotic units are employed. Most of these units are of the shuttle type, where workholders/workpieces are loaded and unloaded by a robotic arm traversing a gantry between the machine and stacked, rotary storage magazines. These are produced by the traditional EDM tooling manufacturers, while the fully articulated robotic arms continue to be manufactured by the companies involved in robotics and work-handling equipment. These types are usually free-standing robotic "arms" that can be moved to the appropriate machine and programmed separately or integrated with the EDM program and operated by command from the machine's control. Some of these have been equipped with vision systems so the robot can "see" the part's orientation and execute a successful pick-up and transfer.

Figure 1.4   An automatic tool changer allows extended hours of unattended machining. (Courtesy of System 3R, USA, Inc.)

For unattended operations of vertical EDM, an automatic tool changer is most commonly used. Like machining centers, this allows multiple electrodes to be used and changed as needed. Some machines have the ability to use a robotic claw or "gripper" in the tool changer and can be used to change to a different workpiece before being returned to the magazine and exchanged for an electrode.

Figure 1.5   Automatic workpiece changers further increase machine autonomy. (Courtesy of Erowa Technology, Inc.)

Another capability CNC EDM offers to unattended machining is discharge dressing of electrodes. Without going into technical detail, discharge dressing is equivalent to a conventional machining center determining when a cutter is dull, having the ability to resharpen it *in the machine unattended*, and, after adding cutter com-

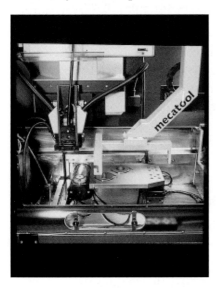

Figure 1.6   A wire EDM pallet that has been automatically loaded onto the machine tool. (Courtesy Mecatool USA, Ltd.)

pensation for the new cutter dimension, resuming machining. Depending upon the application, certain users have not required an automatic tool changer, relying upon redressing a single electrode time and time again.

Like the wire EDM, sinkers also use pallets and pallet changers. These too are typically part of a line of machines in a cell, used to produce the same part or parts, over and over.

PROGRAMMING

Since all wire-cut EDM's are computer numerically controlled (CNC), a computer program must be generated. As part programming becomes more and more complex, the requirement of a good programming system is essential as all but the simplest of programs would prove much too tedious to be entered via the keyboard as manual data input (MDI).

Communicating with this equipment has improved through the years as programmers and software writers have grown to understand the capabilities of the process. Ease of programming is one of the major selling features cited by most manufacturers. All have the ability to communicate with other devices, such as programming stations, printers, plotters, tape punches, etc., through RS-232C serial ports. Some even have internal postprocessors that can obey preprogrammed conversational commands in addition to standard G-codes. Many machines offer on-board programming so operators can program and edit the job at the machine, and almost all makes are compatible with after-market software through postprocessors. Some have DNC capability and can be operated from a remote PC. "User friendly" is a much overused buzzword these days, but in all reality, wire EDM's really and truly are.

CNC vertical machines differ in programming requirements because they typically cut cavities in a Z-minus movement using canned or programmed orbiting cycles, so programming of vertical machines doesn't usually require an external programming system. Except in contouring or "stick-milling" applications, cutter-path programs are seldom needed for vertical EDM.

Most EDM builders supply machines with on-board software to enable an operator to program at the machine. Most of the function of this type of software is to assist in selection and arrangement of cutting technology and orbit offsets rather than generating program paths.

CNC sinkers also share the ease of communication that wire machines enjoy, although they are used far less in this manner.

MACHINE AND TECHNOLOGY TRENDS

Future trends for manual EDM's will be limited as the machine tool is already well designed and built. Any future enhancements will be made on the control and power supply side or perhaps in the servo-system.

All types of CNC EDM machine tools will probably continue to see the integration of the machine, control, power supply, and dielectric system, yielding a smaller "footprint" and more efficient utilization of valuable factory floorspace. The traditional free-standing controls and power supplies are on the wane, as we will continue to see more modular construction with pendant-type controls suspended above the floor or built into panels on the machine itself.

There is considerable activity in the filtration areas by both the manufacturers and after-market suppliers. This area of dielectric purification and waste disposal is reaching volatile stages as environmental concerns in the U.S. continue to increase. The days when an operator can routinely discard used filters into the dumpster will soon be over, and in many areas of the country, it already is.

Power supplies will continue to improve as circuitry gets smaller, yet more powerful. Striving for efficiency by using less power and higher frequencies seems a logical direction considering the sensitivity to recast and surface integrity shown by the aircraft and aerospace industries. Special wire circuitry has already been introduced to eliminate electrolysis and matrix depletion of carbides and other sensitive workpiece materials.

The general consensus among industry spokespeople is that cutting speeds of wire machines will continue to increase, perhaps approaching 40 sq. in./hr. in tool steel in the next few years. At that point they seem to feel that any limitations in speed will be found in the wire itself.

Accuracies and finishes likewise will continue to improve, with finer finishes being obtained with fewer skim cuts. Special circuitry already has finishes down into the submicron range, so it's hard to imagine just how much finer these can get.

Considering the sometimes huge amounts of cutting technology required and the constantly changing conditions faced by EDM users, the machine's controls and computer systems will continue to develop greater and greater capabilities, perhaps ultimately evolving into Artificial Intelligence (AI) units. "Fuzzy logic" enhanced controls (see Chapter 18) are already in use so this trend already has inertia in the direction of machine intelligence. Simply put, these units will monitor and store all previous data on conditions and parameters of what worked and what didn't work. Based upon the data of past performance, it can select the most efficient parameters and continue to improve its future performance.

If all of this sounds like 21st-century concepts, why not? The 21st-century is only a few years away!

MARKET TRENDS

Trends in the EDM market are difficult, if not impossible, to predict; but the EDM market as a whole — and especially that of wire and CNC EDM — will, in my opinion, continue to grow as the technology improves and the applications base broadens. Although the U.S. market is potentially much larger than the markets of Europe and Japan, it is growing at a much slower pace. This is difficult to understand since the spectrum of American manufacturing is so broad and diverse and is, without question, capable of producing anything the Japanese or Europeans are making, and then some. What could be the reason?

Let's compare Japan's EDM purchases to those of the United States. We will examine the number of units from the early 1990's — otherwise, Japan's recent recession and temporary slump in machine tool sales will skew our comparison numbers.

There were approximately 4,550 CNC EDM units installed in Japan in 1991, with over 3,100 of them being wire EDM's. In contrast, the entire U.S. market installed slightly over 1,000 units, down from approximately 1,200 installations in each of the previous two years. Out of 1,000 or so total units, 774 of them were wire EDM's. Both countries show an approximate mix of 70% wire and 30% vertical EDM, so the breakdown of machines is similar but the volume is quite a bit different. What can explain the installation of over *four times* as many units in Japan — an area roughly the size of California?

My guess is *education* and *awareness*.

While EDM has totally revolutionized the mold- and die-making industries, here in America, the other manufacturing disciplines are only just beginning to fully examine or implement this exciting process. Generally speaking, American manufacturing is not fully aware of EDM's advantages over conventional methods. They are learning, but still have a long way to go before matching Japan or Europe's utilization.

At present, EDM use in America is undergoing a huge and rapid diversification, transforming existing manufacturing methods and creating new ones. No longer is EDM use limited to the toolroom. They are seen throughout almost every industry, producing everything from auto and aerospace parts to bone implants and fine surgical devices. Partmaking ranges from R&D and prototyping to high-volume production and, of course, toolmaking.

In 1992, Japan projected a total of 7,000 CNC EDM units to be installed, with almost 4,500 of them being wire EDM's. Although the wire/sinker ratio remains consistent, the volume of total units continues to grow and outpace American consumption for these products. Why is there such a large disparity in numbers and density?

Since Europe and Japan are where most wire EDM's are manufactured, perhaps by "growing up with the process," their engineers are much "closer" to it. They seem to be much more aware of the potentials and capabilities this process has to offer than American manufacturing is. They typically "design for the process," rather than examine EDM only as a last resort.

If this is the case, then acceptance of the EDM process, and a commitment to use it, can only happen:

1) by educating the designers, engineers, managers and other decision-makers in American manufacturing of the total system benefits this process can provide,

2) when American manufacturing is made aware that EDM is not a "black art" or "mystery area" to lose jobs and money into, and

3) when American manufacturing becomes aware of the increased production, reduced secondary operations, unattended operations, etc., that EDM offers.

In the U.S., the main buyers and users of this technology are found primarily in the following standard industrial classification (SIC — also see charts in Appendix B) categories. These are the industries that have the most to gain by accepting and committing to the EDM process.

SIC LEGEND

SIC 2500   Furniture and fixtures
SIC 3400   Fabricated metal products, except machinery and transportation equipment
SIC 3500   Machinery, except electrical
SIC 3600   Electrical and electronic machinery, equipment and supplies
SIC 3700   Transportation equipment
SIC 3800   Measuring, analyzing and controlling instruments

SUMMARY

Most manufacturing applications for EDM already exist, but they are merely waiting to be discovered and implemented. As this occurs, the increased use of EDM in manufacturing will continue to grow and diversify at an accelerated rate through a combination of *necessity* and *imagination.*

Necessity will drive it because as newer and more exotic materials are developed, conventional machining operations will continue to reach their limitations. Necessity has already prompted several manufacturers into successfully EDMing conductive ceramics, alumina-substrate materials, germanium, and certain composite materials simply because there was no other practical means to cut them. Polycrystalline diamonds (PCD's) have also been successfully wire-cut, and technology has now been developed to cut single-crystal diamonds. In manufacturing, there will always be a need or necessity to "find a better way" to make something. EDM has a great deal to offer in this quest.

Imagination will fuel the process by providing the ability to design tools and parts that previously weren't possible or cost-effective by any other method. The prospect of machining complex shapes in hardened or exotic materials to burr-free splint-tenth accuracies will continue to attract engineers and designers who are seeking an alternative method to expensive, more restrictive, and limited "traditional" methods.

Before EDM can gain universal acceptance in American manufacturing, more designers and engineers will have to "take their blinders off" and, with some imagination, gain some "peripheral vision." Very few new methods will be seen from a "head on" perspective, and the most successful manufacturers of tomorrow will also be the most creative. Imagination will allow thinking and planning to make the necessary transition from "traditional" to "nontraditional."

Necessity will *drive* the EDM market, and *imagination* will *fuel* it.

If necessity and imagination are the *drive* and *fuel* for the growth of the EDM industry, then new and continually improving EDM technology will be the *engine* to propel this versatile manufacturing process ahead and on into the 21ˢᵗ -century.

Chapter 2

BASIC EDM THEORY

INTRODUCTION

EDM — the very mention of EDM brings shivers to the spines of the uninformed, and knowing smiles to the knowledgeable. Although the EDM process has been in use for decades, it is still widely misunderstood by many in the manufacturing community. EDM is the process of electrically removing material from any conductive workpiece. This is achieved by applying high-frequency pulsed, ac, or dc current to the workpiece through an electrode or wire, which melts and vaporizes the workpiece material. Positioned very precisely near the workpiece, the electrode never touches the workpiece but discharges its potential current through an insulating dielectric fluid (water or oil) across a very small spark gap. The spark is reported to be in the range of 8000 to 12000°C (14432 to 21632°F), and it vaporizes and melts the workpiece material. This process is used when the workpiece material is too hard, or the shape or location of the detail cannot easily be conventionally machined. This makes many formerly difficult projects more practical, and many times it can be the only feasible way to machine a part or material.

EDM was first implemented over 40 years ago, and used primarily to remove broken taps and drills from expensive parts. These were quite crude in construction with hand-fed electrodes. During World War II, two Russian scientists, B.R. and N.I. Lazarenko, adapted the first servo-system to an EDM machine. This offered some semblance of the degree of control that is required for efficient but safe EDM machining today.

Through the years, the machines have improved drastically — progressing from RC (resistor capacitance or relaxation circuit) power supplies and vacuum tubes to solid-state transistors with nanosecond pulsing. From hand-fed electrodes to modern CNC-controlled simultaneous six-axes machining. And now, augmented by its younger brother, the wire-cut EDM, these two have combined to revolution-

Figure 2.1   Spark sequence. (Courtesy of Agie, USA, Inc.)

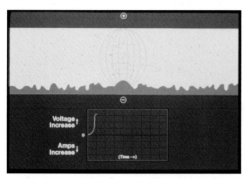

Figure 2.1a   "Open-gap" voltage. The electrode is seeking the workpiece while "cutting air." Graph shows high potential voltage only, and no current. Time-line runs horizontally.

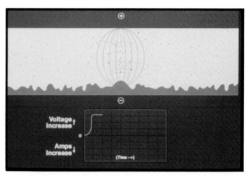

Figure 2.1b   Displays the electromagnetic field created between electrode and workpiece. Dielectric within this field becomes polarized as resistance decreases. Voltage levels off.

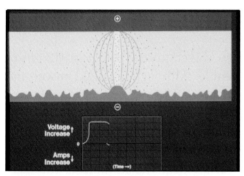

Figure 2.1c   "On-time" begins. Dielectric resistance is overcome and spark occurs, generating current which vaporized the workpiece. As amperage increases, voltage will decrease.

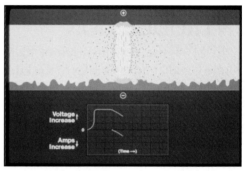

Figure 2.1d   Spark is plasma hot and enclosed within a sheath of gases. Vaporization of workpiece continues.

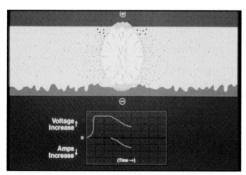

Figure 2.1e   Gas bubble continues to expand rapidly (vapor pressure). At a certain point, vaporization will cease and melting begins. Dielectric contamination increases.

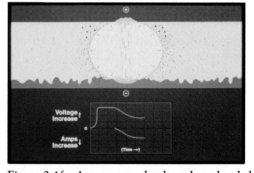

Figure 2.1f   Amperage and voltage have leveled off as contamination and thermal damage of dielectric increases. Dielectric is now severely compromised and its electrical resistivity continues to rise. If allowed to continue, conditions will cause "dc arcing" or a wire break.

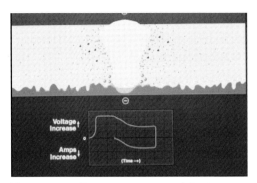

Figure 2.1g   Power is interrupted during "off-time" part of the EDM cycle. Current drops to zero. Gas bubble collapses upon removal of heat source.

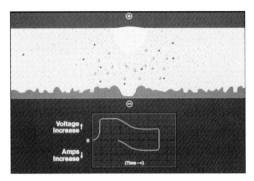

Figure 2.1h   Gases and contaminated dielectric will naturally disperse, but providing forced or "sealed" flushing is best and will significantly reduce dielectric recovery time and increase cutting speed.

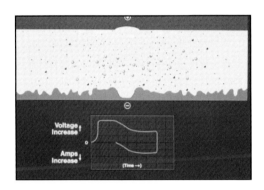

Figure 2.1i   Contaminants and damaged dielectric are expelled, revealing EDM crater on workpiece and wear on electrode. Dielectric begins reionization, allowing repeat of cycle.

ize many industries and to change long-standing and generally accepted processes of manufacturing.

To better understand all this, it is best to start at the beginning.

THEORY

Assuming the electrode is positively charged and the workpiece and table are negative (or vice-versa), the electrode is advanced into the workpiece through an insulating liquid, or *dielectric fluid*. This is usually a paraffin, kerosene, or silicon-based dielectric oil for vertical machines, and deionized water for wire-cut machines. The dielectric fluid is integral to the process. It provides insulation against premature discharging, cools the machined area, and flushes away the "chips."

As the electrode or wire, charged with a high-voltage potential, nears the workpiece, an intense electromagnetic flux or "energy column" is formed and eventually breaks down the insulating properties of the dielectric fluid. Picture the ends of two bar magnets with the north and south poles held apart. If one were to lay a piece of white paper over the magnets and sprinkle fine iron filings onto it, these filings would be caught in this magnetic flux, aligned by polarization in the same way the dielectric is affected.

At this time, with the "grain" of trapped ions in polar alignment and the resistivity of the fluid at its lowest, a single spark is able to flow through this ionized "flux tube" and strike the workpiece. The voltage drops as current is produced, and the spark vaporizes anything in contact with it, including the dielectric fluid, encasing the spark in a sheath of gasses composed of hydrogen, carbon, and various oxides. The area struck by the spark will be vaporized and melted, resulting in a single crater being formed. Due to the heat of the spark and the contaminates being produced by the workpiece, electrode (or wire), and the dielectric fluid, the field of ionized particles becomes "lazy," the alignment is disrupted, and resistivity increases rapidly. Voltage will rise as resistivity increases and the current will drop as the dielectric can no longer sustain a stable spark. At this point, the current must be switched off.

During the time current was flowing through the spark gap, the plasma-hot area was rapidly expanding away from the heat source — the spark. When the current is switched off, there is no more heat source and the sheath of vapor that was around the spark implodes. Its collapse creates a void or vacuum and draws in fresh dielectric fluid to flush away swarf and cool the area. This "off" period allows the reionization process of the dielectric fluid to be completed, and provides favorable conditions for the next spark. The duration of the off-time must be sufficient enough to flush away the spark debris and damaged dielectric, or stability will be difficult to maintain, resulting in dc arcing or a broken wire. This briefly describes one EDM cycle; and in order to machine with this process, it must be repeated over and over again, switching on and off thousands of times a second.

Together, the "on" and "off" pulses (in units of microseconds, or μsec) comprise a single cycle of electrical discharge machining. The length and duration of these parameters will depend upon the material, electrode material, flushing, speed required, and surface finish. Generally speaking, low frequencies are used for rough machining and high frequencies are used for finish machining. Some materials — due to density, conductivity, and/or melting temperature — must be machined with higher frequencies even during roughing operations (titanium, carbide, copper, etc.). This will result in improved finishes and surface integrity, but with substantially increased electrode wear.

The relationship of the on-time to the off-time is the measure of efficiency, better known as the "duty cycle." This is calculated by adding the on-time and the off-time together and dividing this total into the on-time. Multiply this quotient by 100 for the percentage of efficiency, or duty cycle. See below for examples.

Duty cycle = on-time/total cycle time × 100

<u>Examples:</u>
25 on / 25 off = 25/50 = 0.50 × 100 = 50%
75 on / 10 off = 75/85 = 0.88 × 100 = 88%

Obviously, it would be desirous to reduce the off-time to the smallest possible increment, but many variables such as flushing conditions, electrode material, workpiece material, dielectric condition, etc., can drastically affect the ability to maintain concurrent efficiency and stability.

Many of the modern power supplies have the ability to monitor and change conditions and duration of the spark by using "adaptive" controls. This ability can compensate for marginal cutting conditions automatically, even allowing unattended operations.

MORE ABOUT ON- AND OFF-TIME

Knowing how EDM works isn't enough. We need to make it work *for us*. We need to know how to get good machining speed and good finishes with minimum wear and the lowest possible chances of dc arcing or wire breaks. We'll need to examine this in depth because on- and off-time are much more than just a switching cycle. Together, they control the basic parameters of the EDM process.

ON-TIME

Speed: *All* of the work is done during on-time. The spark gap is bridged, current is generated, and work is being accomplished. The longer the spark is sustained (the higher the duty cycle), the more workpiece material will be melted away.

Finish: Consequently, with a longer period of spark duration, the resulting craters will be broader and deeper; therefore, the surface finish will be rougher. Obversely, with a shorter spark duration, the finishes will be finer. Later in this chapter we will discuss the related conditions of spark frequencies and surface finish in the section titled "The Recast Layer."

Wear: With positively charged electrodes, the spark leaves the electrode and strikes the workpiece, doing most of the thermal damage to the workpiece (hopefully). Except during certain roughing operations using elevated on-times when a phenomenon occurs known as "plating," every spark leaving the electrode can take a microscopic particle with it. More sparks produced within a unit of time will produce proportionately more wear. That is why EDM behaves just the opposite of chip-cutting operations. Roughing electrodes tend to last much longer than finishers. Electrode material will also play a large factor concerning the amount of wear, but will not be discussed here.

OFF-TIME

Speed: While the actual work, or metal removal, is accomplished by the spark only during the on-time, the duration of rest pauses required for reionization of the dielectric can drastically affect the speed of the operation. The longer the period of rest (or off-time), the longer the job will take. Unfortunately, off-time is a necessity and an integral part of the EDM process; but the smaller this increment is, the faster the machining operation will proceed — to a point.

Stability: Just as important as machining speed is to an EDM operation, stability is the key to maintaining this speed. Although increasing the off-time will slow down the process, it can provide the stability required to successfully EDM a given application. If the off-time is insufficient, it will cause erratic cycling and retraction of the advancing servo(s), slowing down the operation more than a less efficient, but stable, off-time would.

Analogy: Would you travel farther in a car that could do 100 mph for only very short intervals, or in a car that could do a steady 55 mph all day?

Wear: An argument has been presented to me, on occasion, claiming that wear is also a factor influenced by off-time. Perhaps this perception is due to this reasoning: if 10% wear is incurred in 1 hour, then by doubling the off-time, the job would take twice as long, then the total wear should be 20% (if the job takes longer, there *must* be more wear).

I beg to differ with this opinion, based upon simple logic and undeniable physics. When the current is switched off, it is exactly that — *off!* Nothing (other than the recovery of the dielectric) is occurring. No work is being produced, nor any wear. If only the off-time is changed, *only the time* required to complete the job will change.

An abstract (but adequate) description of this might be: a certain cavity might require 1,000 individual sparks to complete. If these sparks are only 10 microseconds apart, the job should proceed in a normal fashion. If they are 10 *seconds* apart, the job will take much longer, but with no increase in wear. 1,000 sparks are 1,000 sparks — no matter how far apart in time they are! Therefore, *the duration of off-time will not affect wear, only time*.

There may be some very fine points to discuss with theorists, such as: increased off-time would allow the workpiece temperature to cool more between each pulse (how much can this area cool in a few microseconds?), thus requiring the spark to "work harder" in elevating the impact area to the vaporization/melting temperature; but I have never found a change of off-time *alone* to affect wear characteristics — only the speed and stability.

Minimal off-time is a key to machining speed, but unfortunately a sufficient amount of off-time *is* required to maintain machining stability. Stability is more important than speed; in fact, maintaining any kind of sustained machining speed is practically impossible without cutting stability.

POLARITY

In EDM, polarity describes which side of the spark gap is positive or negative. Polarity can affect speed, finish, wear, and stability.

Vertical machines can use both positive and negative polarity, depending upon the application, but most operations are performed using a positive electrode. Positive polarity will machine more slowly than negative polarity, but is used most of the time to protect the electrode from excessive wear.

Negative polarity is used for high-speed metal removal when using graphite electrodes, and should be used when machining carbides, titanium, and refractory metals using metallic electrodes. Negative electrode polarity used with metallic electrodes is slow, but it is used because no other method is as successful. With graphite electrodes, negative polarity is much faster than positive polarity by as much as 50% or more, but with as much as 30% to 40% electrode wear. This is a good choice for large cavities or with shapes that can easily be redressed or abraded. (For more specific information, see the section titled "The Selection Process" in Chapter 3.)

Wire machines almost always run with negative polarity — that is, the wire is negative and the workpiece is positive. As in vertical applications, metal removal rates are higher using negative polarity, but since the electrode (wire) is constantly renewed, electrode wear is not a consideration. However, if the wire is burned deep enough, usually about 20% of its diameter, it can no longer withstand the tension and will break. Increasing the speed of the wire will reduce the severity of the wire erosion and help eliminate wire breakage, at the small expense of increased wire consumption.

THE RECAST LAYER

One drawback of the EDM process is something called the recast layer, or remelt (see Figure 2.2). This recast layer is an inherent byproduct of the EDM process, and although new technology allows recast to be controlled to a very fine degree, the thermal nature of the EDM process itself makes it impossible to eliminate it entirely.

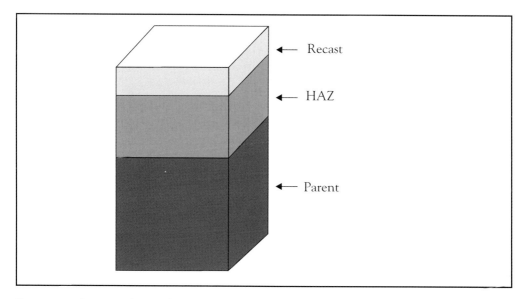

Figure 2.2   Recast and HAZ layers formed by the EDM process.

Here is a brief description of how recast is formed. After a spark has produced a crater and been switched off, a quantity of molten material will be drawn back into the spark crater by surface tension and cooling effects. This molten material will "refreeze" back onto the cooler crater walls. This layer tends to be highly carbonaceous and is called recast material, or the "white layer." Just below the white layer is the area called the "Heat Affected Zone" (or HAZ). This area is only partially affected by the high temperatures. The thickness of the recast layer and the HAZ immediately below it depends upon the current and frequencies used during machining, and the heat-sinking ability of the material itself. Recast can affect the structural and/or surface integrity of the EDMed area.

If the recast layer is too thick or is not reduced or removed by polishing, abrasive flow machining (AFM), electrical chemical machining (ECM), etc., this very hard and brittle recast layer can cause cracking, flaking, impart stress risers and ultimately it can lead to premature failure of the part. In severe cases, recast removal and/or stress relieving of the part is sometimes necessary. This has been a major concern of the aerospace/aircraft industries for years and, only recently, with the improvement in modern power supplies, has acceptance of EDMed parts for aircraft and aerospace industries become routine.

While on the subject of recast and finishes, there are distinct differences between the recast layer left by wire EDM and the recast left by vertical EDM. This is attributed to the different dielectric fluids used, which are usually oil for vertical machines and water for wire-cut. (Sinkers have been modified to use water dielectrics, and a production model has recently been reintroduced to the United States market. Wire-cut machines using oil and oil/water dielectrics are also available, and can provide improved surface finishes, but with a significant reduction in cutting speeds.)

Oil dielectrics change the parent material by producing an uncontrolled heat-treating process — heating the metal to a very high temperature and then quenching it in oil. The high heat breaks down the oil into hydrocarbons, tars, and resins. The molten metal draws the carbon atoms from the oil and they become trapped within the recast layer, creating a "carburized" surface. While this is a far cry from the carefully produced carburized surfaces obtained during professional heat-treating operations, it is nevertheless hard — glass hard, in fact. Even when EDMing prehardened materials, the recast surface produced in oil will generally be several points higher in hardness than the parent material. Ask anyone who has stone-polished an EDMed cavity just how hard the recast or "white layer" is.

Recast produced in water dielectrics is different from oil because oxides produced by the vaporizing water, along with electrolysis, can deplete carbon from the material's surface. In addition, copper atoms freed from the brass wire can be assimilated into the exposed surface of the workpiece material, further contributing to an alloying process and softening the parent material. In ferrous materials, wire-cut surfaces can be several points lower in hardness than the parent material.

Carbides also can be EDM machined, but improper power settings will deplete the cobalt binder material that holds the tungsten particles in place, resulting in flaking or cracking. Fortunately, most modern power supplies are capable of reducing most of this concern by controlling the thickness of the recast material through constant monitoring and control of the spark/servo conditions during high-frequency machining. Another reason for the improvements in the surface integrity of EDMed carbides, especially in newer wire-cut machines, is the absence of capacitors. This will be covered in detail in Chapter 14: Carbide and the EDM Process.

MACHINING TIPS

The following recommendations are guidelines only. Your single best source of cutting technology is from the applications engineers employed by the builder of your machine.

For Speed: As usual, EDM is always a fine line of compromise. In this case, it is speed versus stability. If the overcut, wear, and finish are satisfactory and flushing is adequate, the machining speed of an EDM operation can be increased by slowly decreasing the off-time in small increments (1 - 5 μsec), until machining becomes erratic, and then quickly returning to the last stable setting. Now, this isn't a very technical-sounding solution, nor does it require a lot of formulas, but not many of us are mathematicians, nor do we desire to be. We just need results in a hurry, and this method works.

As you do this, the machining or gap voltage will slowly decrease as the working current (amps) will slowly increase. If the room is quiet enough, you may also be able to actually hear the discharge frequency change as you increase or decrease the on- or off-time. Try not to let the gap voltage drop below 35 to 40 V as this can be risky even with optimum flushing; in fact, many times gap voltages can't be set lower than 100 V without difficulty. Of course, all of this depends upon your application, flushing conditions, electrode selection, servo voltage, and other power supply settings.

For Wear: Wear is not a serious consideration for wire EDM because of variable wire speeds and the constantly renewable electrode material from the spool. For vertical EDM, leaving a minimal amount of stock for finishing will also help keep electrode wear to a minimum. Using no- or low-wear settings to remove most of the material and leaving only the smallest amount of stock for "changing the finish" is the correct method.

Logic: If there is only a few thousandths (0.003 to 0.005") of stock left for finishing operations, even at 20% wear, how much could this possibly amount to? (20% of 0.005" = 0.001"). If the tolerance is less than this amount, redress and go in again (20% of 0.001" = 0.0002").

For Finish: In EDM, surface finishing is usually the most time consuming, therefore, the most expensive. Careful planning and preparation can offset this. For both wire-cut and vertical CNC machines, careful selection of the current and frequencies used and the amount of material left to remove is paramount. Whether you are skim-cutting with wire or orbiting with ram, the remaining amount of material for each step should be only slightly more than the maximum crater depth left by the previous cut, otherwise you can wind up "chasing" old finishes and heat affected zones.

The old shop saying, "Rough it out to size," almost applies here, as long as the heat from the roughing operations does not penetrate too deeply into the workpiece and there is sufficient stock remaining to remove all traces of the previous finish.

When attempting parts or details requiring minimal recast, or "mirror finishes," you should not be concerned with metal removal in the final steps, as this is the duty of the roughing and semi-finishing operations. At this point, your part or cavity should be only a few tenths away from its finished dimensions, so we want only to change the surface finish, actually removing very little material. Using low power, very high frequencies, and the minimal amount of offset for each cut is the best strategy.

Fine finishing on manual machines is more difficult because of the inability to orbit the ram or X/Y-axes. Most jobs on manual machines are finished with several passes of a slightly larger finishing electrode. The problems associated with this operation are: the time required for each finishing "pass," uneven electrode wear, and the time and support equipment used to make or redress multiple electrodes. The reason for the difficulty in this type of operation is that the very small area remaining for finishing (usually 0.005 - 0.010" per side) cannot accept very much current. In addition, while using finishing conditions (wear settings), excessive wear of the leading edges of the electrode occurs because it must cut the *entire Z-depth again* to finish the side walls, instead of just the few thousandths of material actually remaining in the X- and Y-planes. So, even though the cavity has been "finished," it usually isn't. The worn electrode must be redressed or replaced by a new one to remove the material left by the first finisher. In many cases, several dressings and reentries must be made before the electrode "sparks out" and side-wall taper and corner radii are within blueprint requirements.

The speed and performance of finishing operations on manual machines can be greatly improved at a relatively small cost, by adding an after-market orbiting device to the head or ram. While an orbiting device will not make a manual machine into a CNC, it will help improve speeds, finishes, flushing, and reduce electrode corner wear. Orbiting is covered in detail in Chapter 5.

Mirror finishes are different in their approach and electrode preparation, and are seldom obtained without motion during machining, whether by orbiting or electrode rotation. Longer machining times and higher wear are unavoidable and should be factored into the price of the job.

We have covered the basics of EDM theory, on-time, off-time, duty cycles, polarity, and recast. In the following section, we will examine the roles of current and frequencies, their effects upon finish and surface integrity, and how to meet our requirements of speed, wear, and finish.

MACHINING PARAMETERS

After understanding the basic roles of "on-time" and "off-time," the other main factors of metal removal by the EDM process are: current and frequency. Each is distinctly different but must be used together in different combinations to obtain the desired results.

CURRENT

This is the amount of power used in discharge machining, measured in units of amperage. In both vertical and wire applications, the maximum amount of amperage is governed by the surface area of the "cut" — the greater the amount of surface area, the more power or amperage that can be applied. Higher amperage is used in roughing operations and in cavities or details with large surface areas.

For vertical applications, the "rule-of-thumb" for maximum power selection is approximately 65 amps per square inch of electrode engagement. This simple formula would be based upon the surface area (SA) of electrode engagement multiplied by the constant of 65 amps per square inch, or

$$SA \times 65 = maximum\ amperage.$$

For example, the maximum amperage for an electrode ½-inch square would be calculated by multiplying the length of one side times the length of the other and then multiplying this number by 65 (0.5" × 0.5" = 0.25" × 65 = 16.25 A). This figure is conservative for smaller sizes since improved materials, especially graphites, can accept higher currents without excessive damage.

This general rule applies to copper as well as graphite electrodes, although wear for each material will vary depending upon the on-time. This will be covered in detail in Chapter 3.

Although each square inch of electrode surface can withstand 65 amps, this maximum is seldom approached except in the largest of surface areas. While the electrode itself might withstand this intense current, the workpiece, depending upon the material, can suffer from excessive heat, and the depth of recast material may not entirely "clean up" during finishing operations. This involves more than just "cleaning up" actual crater depth and recast. In high amperage/low frequency roughing operations, the intense heat that is generated can sink deeply into the area surrounding the actual burn and this material can undergo an uncontrolled heat-treating or annealing process, depending upon the material. In severe cases, this damage can penetrate deeper than finishing operations will remove. Also, when using maximum amperages, the spark gap can be quite large, and smaller details may have to be omitted from the electrode and done later with less power.

The chart shown in Figure 2.3 is actual size and can be used as a reference for calculating actual amperage per square inch, or it can be used as a "cheat sheet." This chart can be duplicated and laminated for protection and used in the shop. By merely placing the electrode over the appropriate-sized square or rectangle, am-

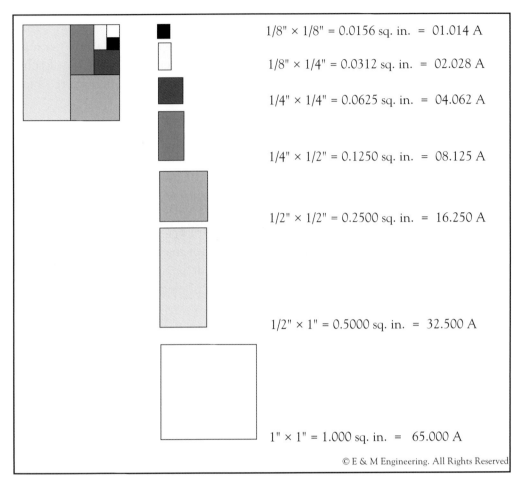

1/8" × 1/8" = 0.0156 sq. in. = 01.014 A

1/8" × 1/4" = 0.0312 sq. in. = 02.028 A

1/4" × 1/4" = 0.0625 sq. in. = 04.062 A

1/4" × 1/2" = 0.1250 sq. in. = 08.125 A

1/2" × 1/2" = 0.2500 sq. in. = 16.250 A

1/2" × 1" = 0.5000 sq. in. = 32.500 A

1" × 1" = 1.000 sq. in. = 65.000 A

Figure 2.3   Electrode "cheat" sheet.

perage can quickly be "guesstimated." By using this chart as a guideline, diagrams displaying larger or circular areas can be made.

The power settings shown on this chart are conservative in the smaller sizes and closer to actual maximums for sizes 0.250 sq. in. and larger.

The actual overcut will be determined by the amount of current and the on-time; but usually when elevated currents are applied, it is with an undersized electrode which will leave sufficient material to be removed later in subsequent finishing modes using less power and orbiting, or by changing to larger, finishing electrodes.

FREQUENCY

This is the measure used to gauge the number of times the current is switched on and off. During roughing operations, the on-time is usually extended for high rates of metal removal; and since there are fewer of these cycles per second, this would be a low-frequency setting. Finishing operations, with much shorter on- and off-times, will have many cycles per second and would be considered high

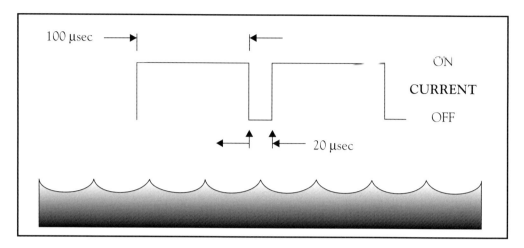

Figure 2.4   Low-frequency/roughing.

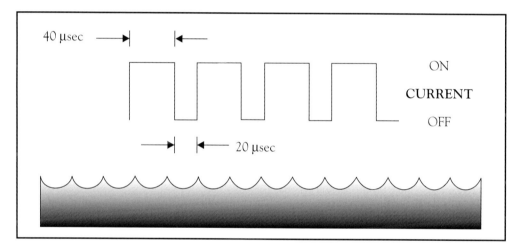

Figure 2.5   Moderate-frequency/semi-finishing.

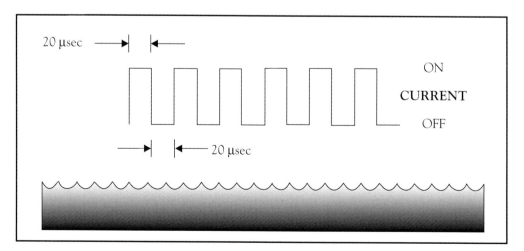

Figure 2.6   High-frequency/finishing.

frequencies. Many times you can actually hear an audible frequency change when changing the on- or off-time while a burn is in progress. Frequency is not to be confused with duty cycle which is a measure of efficiency. Figs. 2.4 through 2.6 portray EDM wave forms (much simplified) at different frequencies, and the representative effects upon finish.

As you can see in Figure 2.4, the finish left by long on-times is quite rough. This is because the long duration of the spark is sufficient enough to allow a great deal of heat to sink into the workpiece, melting a large crater, rather than vaporizing a small one. In addition, the recast layer will be considerably thicker with a potentially deep Heat Affected Zone, or HAZ. This can present problems with the surface integrity of the part unless sufficient stock is left for removal by secondary finishing operations, using EDM, ECM (Electrical Chemical Machining), AFM (Abrasive Flow Machining), or manual polishing.

In Figure 2.5, the spark duration has been decreased, producing smaller craters and less recast. The finish is improved, but machining speed will drop off and, in the case of vertical EDM, electrode wear will increase.

Figure 2.6 demonstrates the wave form of a finishing operation. As you can see, the spark signature is much more dense, with many more sparks within the same unit of time. With sparks of short duration, there is very little material removed, as you can see by the smaller craters. This is the method used in finishing EDMed cavities; by lowering the power and on-time until the desired size and finish are obtained. There is also much less potential of thermal damage to the workpiece using high frequencies. That is why carbides, which are very susceptible to surface cracking, are machined with high frequencies.

SUMMARY

EDM is like the ancient game of "Go" or its modern counterpart "Othello," in the sense that while it is relatively easy to learn, it is difficult to master. My personal observation from having used and taught EDM for many years is that it can be equated with learning the English language: "This rule will apply *every single time, without fail*... except, of course, for the 4,368 exceptions that you may occasionally encounter."

Do you remember that one? I try not to.

All kidding aside, there are times when EDM can be so frustrating you could pull your hair out; and other times it can be so simple that it is elegant. It is, and always has been, a tradeoff — a compromise — accuracy versus speed, speed versus finish, finish versus wear, etc.

Those of us who use, or are around, the EDM process are very lucky to be involved with such a fascinating and potentially profitable technology. Even though the process is over fifty years old, it is still considered to be an emerging technology because improvements occur at such a rapid pace. Just think — wire cutting speeds are almost *thirty* times faster than they were just fifteen years ago! The accuracies and mirror finishing capabilities of the modern vertical machines are certainly a far cry from the crude "tap-busters" of just a few decades ago.

EDM theory is just that — theory. It all looks good on paper, but it doesn't necessarily apply on the shop floor. EDM theory is the foundation for this process's role in modern manufacturing. What actually happens on the shop floor is up to the operator, the manager, and the designer. Those who have the abilities and imaginations to solve unique problems, or those who have the vision to plan around them, are the keystones of this industry and are good for it. They are usually the ones demanding that new model machines be equipped with *this* widget or *that* gadget. They are the ones wanting to see *this* type of tooling or *that* type of wire. Ultimately, the entire manufacturing community profits from these "free thinkers" and their abilities to improvise to do the "impossible" jobs, and pressuring manufacturers and suppliers to continually improve their products.

In closing, once again, theory is just theory. What is done with it on the shop floor is what is important. The successful integration of theory into practice is the engine that runs this industry.

Keep up the good work!

SECTION II

VERTICAL EDM

Chapter 3

ELECTRODE SELECTION AND FABRICATION

INTRODUCTION

Electrode selection is only one of the many facets of vertical EDM that makes learning and using it successfully much more difficult than its younger counterpart, wire-cut EDM. While users of wire-cut machines do have a selection process to complete, there is no fabrication involved. Their "electrode" is store-bought and comes wound on spools and is designed to be expendable. The vertical EDM operator faces quite the opposite challenge. He or she must select the proper electrode material and then, while "designing around flushing," must choose the best way to fabricate it, depending upon the material. Then, in almost every case, the operator must strive to protect this electrode from excessive wear, while maintaining efficient cutting speed. Very seldom will all of these be an easy task. Let's examine why.

MATERIAL SELECTION

This is where we can generate no small debate. We have a choice between two popular types of electrode materials — copper and graphite. Some proponents of each material stand almost immovable when it comes to using other than the "right" choice — theirs! The disappointing news to both sides of this debate is that none is universally "right." Each material has its own unique properties and will fit certain applications better than others. Some are used correctly, and some are used in their absolute worst applications. We will try to shed some light on this argument that is almost as old as the EDM process itself.

When EDM first became a reality almost fifty years ago, it was a very crude process used to "bust taps" or remove broken drills, bolts, etc., from expensive or hardened parts. Two Russian scientists improved upon this with the addition of a servo system, thus the first "true" EDM machine came about. In Europe, as well as Japan, their primary choice for electrode material was (and still is) copper and its alloys. In part, because metallic electrodes were used first (actually, graphite as an electrode material had not yet been considered), and when graphite *did* become available, it was "too dirty" to be easily accepted.

Initially, brass was the only electrode material. It was readily available and inexpensive, despite its high rate of wear. Later, the logical choice led to the use of copper and its alloys. They offered higher conductivity and much improved work-to-wear ratios over brass. As improvements in equipment made EDM more feasible for "everyday" applications, the physical limitations of straight coppers became evident. The element copper melts at approximately 1083°C (1980°F), while temperatures in the spark gap must exceed 10000°C (18064°F). With EDM spark temperatures many times higher than copper's melting point, certain applications using copper electrodes were no longer cost effective due to unacceptable wear.

With the advent of graphite as an electrode material, a far superior Metal Removal Rate (MRR) was attained, and with far less wear. Both of these advantages are realized through graphite's ability to resist thermal damage while discharge machining. Graphite is considered a "metalloid." It does not melt like most metals, but undergoes a process called "sublimation." That is, it goes directly from a solid to a gas, bypassing the liquid state entirely. It does this at approximately 3500°C (6332°F). This is the main reason graphite resists wear better than copper — its superior resistance to thermal damage.

In addition to providing superior wear resistance, ease of fabrication is another plus for graphite. It is far easier to machine or grind, with no resulting burrs to remove, as is the case with copper. In milling operations, copper throws a substantial burr on all the "off" edges. Unfortunately this is unavoidable. It also poses problems in grinding operations. Depth of cut is severely limited and the wheel must be charged with beeswax or something similar to prevent excessive loading of the wheel. Wheels with "open" grain structures (46-J) should be used when flat-grinding copper. This provides large openings to contain a large "charge" of wax and lots of room for the soft gummy copper chips. The downside of this is that the leading edges of the wheel will break down rapidly, requiring frequent redressing. Grinding wheels of at least 60 and sometimes 80 grit are necessary when grinding copper electrodes with sharp corners and fine detail. These will consequently cut "hot" and load up much faster, but using the harder wheel is the only way to keep sharp corners and fine details from breaking down too rapidly.

While on the subject of redressing, copper holds a large advantage over graphite by its ability to be discharge dressed. Discharge dressing is the process whereby we "trick" the EDM machine into wearing out the electrode and protecting the copper tungsten dressing block (workpiece). This can be done manually (usually in only the Z-axis), or by CNC (any/all axes), although it is much more efficient in a CNC machine. Copper will discharge dress much faster and impart far less wear to the dressing block than will graphite electrodes. This is due to the same reasons graphite has less wear in machining operations. It has greater resistance to spark erosion, therefore it will discharge dress much more slowly. So in CNC EDM operations, when a tool changer is *not* used, unattended or overnight operations should run faster with copper electrodes because they will discharge dress much faster than graphite.

If difficulty in machining and burrs are negative aspects of copper, then the clouds of graphite dust (and the wrath of fellow workers) generated by machining graphite is definitely a negative. One must consider the cost of an efficient vacuum system to control this problem if graphite is a consideration at all. While this dust is not actually toxic, it can cause respiratory problems and allergic reactions, especially the graphites infiltrated with copper. In general, all graphites are a housekeeping nuisance, for both personnel and equipment.

While there are powdered graphites that are used as dry lubricants, the graphites used for electrode materials are synthetic and very abrasive. That is why carbide-tipped tools and carbide end mills are recommended in graphite electrode fabrication. The graphite material does not "cut." That is to say, it does not shear and flow across the tool edge like metals do. It fractures, or crushes, under the cutting tool pressures, and floats away as a fine powder or dust. Under a microscope, these particles appear crystalline in structure and are very sharp. When these particles mix with the film of oil on machine ways, this makes an abrasive slurry that will cause premature wear and decrease the machine tool's life and accuracy.

Another consideration is raw material costs. A premium grade of graphite is, on average, three times more expensive than copper. Logic would indicate purchase of the less expensive material, but one must consider the complexity or degree of difficulty in fabricating and/or redressing these electrodes. Despite an initial cost savings of three times for raw materials, if the electrode shape is complex, then fabrication costs by conventional machining (other than wire-cutting) can quickly offset any savings realized by purchasing cheaper materials. So, obviously, electrode selection is seldom simple or easy, especially for entry-level users. All aspects of the EDM job must be examined before electrode material selection can be made.

So far we've discussed only "straight" graphites and copper. There are many alloys and "infiltrated" materials available. We'll look at the metallic group first.

METALLICS

1) **Brass/Zinc**. The first metallic electrodes were made of brass. Due to its high rate of wear, brass is no longer a viable material for electrode use (with the exception of small-diameter tubing for hole-drilling). Zinc or epoxy is used

as a base or filler material that can have a layer of copper electrically deposited from 0.005 to 0.100" thick, and used as an electrode.

2) **Tellurium Copper**. The most commonly used metallic electrode material. The element tellurium has been added (0.5 to 1%) for ease of machining. It has acceptable metal removable rates and reasonable wear. It is very "forgiving" in marginal cutting conditions. It also lends itself very well to wire-cutting and discharge dressing. Sometimes called "Telco."

3) **Copper Tungsten**. Used when the "safety" of copper is desired, but it provides much better wear resistance. Also used in the EDMing of carbides and refractory metals (cobalt, molybdenum, etc.). When mixed and sintered with the element tungsten, copper shows a marked increase in resistance to wear. This is due to the high melting point of tungsten, around 3370°C (6098°F). A tungsten/copper ratio of 70/30 is most commonly stocked by suppliers. It can be readily wire-cut.

4) **Silver Tungsten**. Used where the higher conductivity of silver is desired. As the name suggests, this material is also very expensive, about five times the cost of copper tungsten. It is obviously only for very specific applications.

5) **Tungsten**. While not an "everyday" selection, straight tungsten is used when wear and/or strength is a consideration, not speed. Straight tungsten will EDM very slowly when compared with common metals, but when combined with copper, machining rates will increase. It is very expensive compared with other alloys, and is limited in the sizes available. Due to the difficulty in machining tungsten, electrodes are usually supplied as "pre-forms." Its most common application would be small hole drilling operations (tubing and rod).

EVALUATION OF METALLIC MATERIALS

Advantages: Low cost, high strength, machining "safety," good for entry-level or inexperienced operators, mirror finishes, discharge dressing, wire-cuts readily, "clean."

Disadvantages: Low grindability index, burrs, slower machining speeds, higher wear.

Next, we will examine the graphites. While there are several manufacturers offering many different grades and densities, we will explore only the two basic types of graphites — straight and infiltrated.

GRAPHITES

1) **Straight Graphites**. These are commercially available in many grades from submicron to "Kingsford." The density and the grain size will be a key factor in its performance and cost. If your application does not require good finish or fine detail, then you can obtain satisfactory results with a less expensive graphite. If they are your primary considerations, then a material with small

grain size and high density should be your choice. A general rule in choosing graphite: the higher degree of wear resistance, the better the finish, and the finer the detail — then the higher the grade of graphite required.

2) **Infiltrated Graphites**. A mixture of superfine copper particles in a graphite matrix can be used in specific applications where good machinability of the electrode is desired (a complex shape), but with the "forgiveness" of copper in marginal cutting conditions (deep slots, poor flushing). The greater the percentage of copper, the lower the chance of arcing, due to its higher conductivity. But a proportionately greater degree of wear will be incurred, and the actual MRR will also decrease slightly. Due to its expense, in comparison with straight grades, careful consideration must be given in its selection.

EVALUATION OF GRAPHITE MATERIALS

Advantages: High strength, good machinability, high MRR (speed), excellent wear resistance, can be abraded or ultrasonic machined.

Disadvantages: Cost, lower degree of "safety" in difficult machining conditions, dust from machining, wire-cuts slowly.

COMPARISON OF METALLICS AND GRAPHITES

ADVANTAGES:

Metallics: Low cost, high strength, machining "safety," good for entry-level or inexperienced operators, mirror finishes, discharge dressing, wire-cuts readily, and "clean."

Graphites: High strength, good machinability, high MRR (speed), excellent wear resistance, and can be abraded or ultrasonic machined.

DISADVANTAGES:

Metallics: Low grindability index, burrs, slower machining speeds, *and* higher wear.

Graphites: High cost, lower degree of "safety" in difficult machining conditions, dust from machining, and wire-cuts slowly.

Now that a few facts about each material have been outlined, perhaps your selection of electrode materials will be somewhat easier. Of course, every application is different and must be individually evaluated. Items to be examined are: How much material must be removed, and what is the finish required? What are the raw material costs? How many electrodes will be needed, and how difficult will it be to construct them? What about redressing?

Using the guidelines above will help to answer some of the questions concerning electrode material selection.

RAW MATERIAL CONSUMPTION

As for the question "who uses what?" Figure 3.1 demonstrates the approximate percentage of use of different electrode materials. This is an averaged figure based upon national statistics supplied by several major distributors of raw electrode materials. By no means is this to be interpreted that one material is "better" than another. This chart reflects only their actual percentage of sales, *by order*, not by dollar amounts.

Graphite	85%
Copper	6%
Infiltrated Graphite	4%
Copper Tungsten	3%
Tungsten	1%
Others	1%

Figure 3.1   Approximate percentage of use of electrode materials, ± 2%.

So far we have covered only the basic differences between copper and graphite. Much discussion remains concerning:

What electrode material is correct for what type of work piece material?

What determines the use of positive or negative polarity?

High or low frequencies?

When should capacitance be used?

What about flushing?

These and other variables are to be considered in the selection process, and all are factors in the success or failure of this type of operation. By paying careful attention to electrode material selection and employing good EDM and toolroom practices, success is assured in almost every case. But to simply conclude that one material or the other is the "only" one to use, or that there is a single "universal" electrode material, is limiting one's capability and consequently (and even more importantly) profitability.

THE ELECTRODE SELECTION PROCESS

So far in this chapter, we have examined the basic differences between metallic and graphite electrodes. Now we will take a look at the actual selection process of the electrode material for the given application. We could get bogged down with a great deal of complicated laws of physics, but I will try to keep this discussion as "close to the shop floor" as I can. If I depart from this premise, it will be very briefly, and only to make a point.

The electrode material selection process should start with the consideration of five important factors:

1) **Metal Removal Rate**. This is the measurement of material removed in a given amount of time; most commonly referred to as "cubic inches per hour." The abbreviation is MRR.

2) **Resistance to Wear**. This is one of the most important aspects of electrode material selection. There are four types of wear to consider: volumetric, corner, end, and side wear, with corner wear usually being the greatest concern.

3) **Surface Finish Desired**. The finish required can sometimes dictate the selection of electrode material. For a fine finish using graphite, the electrode material should be dense. "Mirror" finishes are usually obtained with multiple metallic electrodes or graphite electrodes and motion (orbiting or rotation).

4) **Fabrication Costs**. This depends upon the complexity of the electrode. The more complex the electrode is, the higher the fabrication costs will be. The degree of difficulty in fabrication of raw materials, and/or the possible deburring operations, must be evaluated.

5) **Raw Material Costs**. This is usually only a small part of the total EDM time/cost evaluation. Unless it is an unusually large or high volume application, considerations 1 through 4 need to be ascertained before this becomes a factor.

Now that we know the parameters of electrode material selection, the next step is for us to consider some basic facts concerning the nature or properties of the materials that are to be EDMed and, likewise, of the electrode materials from which we are to choose. We must be aware of several things before we even consider the actual detail or application. First we will examine the workpiece.

The approximate melting temperature of the workpiece material should be known. You don't have to know the "exact" temperature at which this occurs, but you should at least have a working knowledge of the material you wish to EDM. With all the different alloys and exotics in use today, this could prove difficult at best, so a good desk reference should be readily available. The one "closest to the shop floor" would be *Machinery's Handbook*, although there are many other excellent reference books available. Knowledge of the approximate melting temperature alone will usually be enough to start with, but it would also be useful to know the material's specific gravity or, in layman's terms, its density. Knowing the specific gravity will give us a good idea of "how many molecules we will have to remove, and if they are they big ones, or small ones." We will examine this later, but first let's discuss melting temperatures.

MELTING TEMPERATURES

The knowledge of the *exact* melting point of the material is not necessary; only the knowledge of the *range* of a similar group of alloys is required. For example, most ferrous metals have a similar melting point, as do most austenitic materials. While there are exceptions to every rule (especially in EDM), for the most part any differences in melting temperatures between similar alloys would be too small to affect our selection process. This applies to other alloy "families" as well, including aluminums, coppers, etc.

Often, only the melting point of the workpiece and the working temperature of the electrode material need to be known. If the alloy has a low melting point — for instance, aluminum — the first choice of electrode material would be copper. Copper's melting temperature is 1082°C (1980°F). When alloyed with tellurium it is only slightly higher at 1095°C (2003°F) maximum, and this temperature is much closer to the melting temperature of aluminum (660°C/1220°F) than graphite's sublimation temperature which is approximately 3500°C (6332°F). While graphite does prove satisfactory in eroding aluminum, it's almost "overkill" to allow such a wide difference in working temperatures (almost 5 times the difference). If you'll forgive the analogy, it's like using a sledgehammer on a thumbtack. It works, but is it necessary?

The chart in Figure 3.2 is a guideline to be used to familiarize yourself with some common workpiece materials and some of their characteristics.

MATERIAL	SPECIFIC GRAVITY	DEGREES FAHRENHEIT	DEGREES CENTIGRADE	CONDUCTIVITY*
Aluminum	2.70	1220	660	63.00
Cobalt	8.71	2696	1480	16.93
Copper	8.89	1980	1082	97.61
Manganese	7.30	2300	1260	15.75
Molybdenum	10.20	4757	2625	17.60
Nickel	8.80	2651	1455	12.89
Carbon Steel	—	2500	1371	12.00
Titanium	4.50	3308	1820	12.73
Tungsten	18.85	6098	3370	14.00

* Conductivity values are based on silver = 100.00.

Figure 3.2   Characteristics of workpiece materials showing specific gravity, melting point, and conductivity.

Based upon melting temperatures alone, the best results will usually be obtained by matching the electrode's melting temperature to that of your workpiece. Low-temperature alloys such as aluminum, brass, and copper should be EDMed with low-temperature metallic electrodes. Higher-temperature alloys such as carbon and austenitic steels will warrant the use of graphites, which have a much higher "working" temperature.

A general "rule of thumb" is: "Low" temp alloys = metallic electrodes
                                  "High" temp alloys = graphite electrodes.

But, of course, this is EDM — and there always seem to be exceptions to the rules. In this case, tungsten, cobalt, and molybdenum are materials with a higher-than-average melting point, and should, in theory, be eroded with a graphite electrode. Not so. The higher frequencies required to successfully EDM these types of materials would wear graphite electrodes at an excessive rate because their sintered structure is so much larger in comparison to metallic materials. The very short on-times required to EDM these materials will not result in as much *latent heat* to damage the metallic alloys with lower melting temperatures. Figure 3.3 is a reference chart to show the most common choices of electrode/workpiece applications and recommended frequencies in use today.

WORKPIECE	ELECTRODE	ROUGHING	FINISHING
Steel	Graphite	Low	Med/High
Stainless	Graphite	Low	Med/High
Aluminum	Cu/CuW	Low	Med/High
Copper	Cu/CuW	High	High
Titanium	Cu/CuW	High	High
Carbide	Cu/CuW	High	High

Figure 3.3   Frequency recommendations. (Cu = Copper. CuW = Copper Tungsten. W is for Wolfram, the element from which tungsten is obtained.)

Copper or copper tungsten electrodes can be used with good results on carbon and stainless steels, but with less efficiency in both MRR and wear than the correct graphite selections. Using metallic electrodes on steels will provide a higher degree of safety when flushing conditions are poor. The advantage of safer EDMing with copper can be integrated with the ease of fabrication that graphite offers by using infiltrated graphites. In addition, copper infiltrated graphites can be successfully used to erode carbides and will usually yield an almost 50% faster MRR than that of copper tungsten, but at approximately *7 times the wear rate*. Careful consideration must be given when weighing the speed of infiltrated graphite against carbide versus the expense of raw materials and fabrication/redressing costs. All things considered, tungsten carbide remains one of the most challenging of EDM tasks today.

While some materials require the use of higher frequencies (titanium, carbides, and refractory metals), these operations can still be relatively efficient. Although, at these higher frequencies, any changes (however small) can substantially affect efficiency. A factor very important to the EDMer is, of course, efficiency. This can be measured by the relationship between the "on" and "off" times. This percentage is called the "duty-cycle" (see Chapter 2 for information on how to determine the duty cycle), and the more efficient a "duty-cycle" is, the less time the job will take (assuming, of course, that stability is maintained).

ELECTRODE WEAR

When predicting electrode wear, several combined parameters will determine the part finish and consequently the amount of wear realized in any given "burn." Primarily, it is the ability of the electrode material itself to resist thermal damage, but the electrode's density, polarity, and the frequencies used are also a major part of the "wear equation."

Figure 3.4   There are dramatic differences in electrode wear between copper and graphite electrodes.

Graphites have a much higher resistance to heat and wear at lower frequencies, but will wear significantly more during high frequency and/or negative polarity applications. The use of high frequencies with graphites is usually reserved for finishing operations with very little material to be removed, or for high-speed negative polarity operations, in which case the electrode is considered "expendable." This might be quickly generating clearance holes or for roughing simple shapes that can be easily fabricated or redressed.

Figure 3.5 outlines some typical EDM applications and the approximate results obtained by using different electrode materials. These figures are general in their scope, allowing for the wide variances in power supply capabilities, but are acceptable references to material selection and electrode durability.

SURFACE FINISH

Without delving too deeply into EDM theory or the phenomena involved during EDM operations, suffice it to say that all finishing operations are done using reduced power settings in conjunction with high frequencies. These parameters will *always* produce more wear to the electrode, *regardless of the electrode material*. The degree of wear will depend upon the type and grade of material and the finish desired.

Wear can be incurred with every single spark that leaves the electrode (except during the phenomena of no-wear or "plating" of the electrode, when using very

ELECTRODE	POLARITY*	WORKPIECE	CORNER WEAR	CAPACITANCE
Copper	+	Steel	2 - 10%	No
Copper	+	Inconel	2 - 10%	No
Copper	+	Aluminum	< 3%	No
Copper	-	Titanium	20 - 40%	Yes
Copper	-	Carbide	35 - 60%	Yes
Copper	-	Copper	35 - 45%	Yes
Copper	-	Copper Tungsten	40 - 60%	Yes
Copper Tungsten	+	Steel	1 - 10%	No
Copper Tungsten	-	Copper	20 - 40%	Yes
Copper Tungsten	-	Copper Tungsten	30 - 50%	Yes
Copper Tungsten	-	Titanium	15 - 25%	Yes
Copper Tungsten	-	Carbide	35 - 50%	Yes
Graphite	+	Steel	< 1%	No
Graphite	-	Steel	30 - 40%	No
Graphite	+	Inconel	< 1%	No
Graphite	-	Inconel	30 - 40%	No
Graphite	+	Aluminum	< 1%	No
Graphite	-	Aluminum	10 - 20%	No
Graphite	-	Titanium	40 - 70%	Yes

* Denotes Electrode Polarity

Figure 3.5   Typical electrode applications and approximate results.

high on-times). Low frequencies provide fewer sparks in a given unit of time — hence low or no wear. Higher frequencies will allow many individual pulses within a given unit of time. The result is:

small sparks = small craters, but the tradeoff is: *more sparks = more wear*.

Graphite, while being far more resistant to damage from heat, is much less dense than copper. It is made from very fine (depending upon the grade) carbon powder and coal tar pitch, and is sintered in a furnace. Even with today's advanced methods of refinement, graphite's smallest grain structure at the time of this writing is slightly less than 1 micron, or 1μ. Since graphites will always have "more air" within their structures, they will always wear more than metallic electrodes during finishing operations.

Copper has a much finer grain structure, since it and its alloys are metals. Compared to the *atoms* of copper (remember, copper is an element), graphite's structure can never approach the density of copper or its alloys. Although copper is more susceptible to thermal damage than graphites, there is very little heat generated in high-frequency operations (finishing); therefore, copper's higher density will yield less wear in high-frequency applications than will graphite electrodes.

When finishing cavities on manual machines, electrode wear for both materials will be higher, but finer finishes are obtained more readily when using metallic electrodes. This also is due to the metal's higher density. Graphite's "open" structure will "imprint" its grain into the cavity bottom — the coarser the grain, the coarser the finish.

In CNC EDM applications, the wear for both materials will decrease significantly, and finishes from graphite electrodes are comparable to metallic electrodes due to the translation or movement of the table during machining. This prevents the more "open" structure of the graphite electrode from "imprinting" into the workpiece. Instead, the movement of the electrode or workpiece will "saw off" or "average out" the size and depth of the electrode's grain size and thus the depth of the craters.

One final consideration concerning finishing operations is dc arcing. As per Murphy's Law, we seldom encounter difficulty until the job is almost finished (and therefore expensive). Only *then*, it seems, will pitting or arcing occur. Naturally, we have little or no stock left to remove, and never enough time to do it over.

The increased chance of dc arcing during finishing is due primarily to the exceedingly small spark gaps usually encountered in fine finish operations. Since the structure of graphite is so much larger in comparison to copper, the particles dislodged from graphite electrodes have less room to be flushed from the gap. Since electricity will take the path of least resistance, the next spark will likely strike this particle. If this particle becomes trapped due to poor flushing, it becomes a "lightning rod," and will attract a continuous discharge of current, and almost instantaneously create an expensive piece of scrap.

A final thought on comparison of use: all things being equal (power settings, flushing, etc.), you will "get into trouble" quicker using graphite than you will using copper.

ELECTRODE FABRICATION

The next thing to look at is the machining or fabrication costs of the electrode. In every case except wire-cut operations, graphite electrodes will take less time to conventionally machine than metallic ones. There is no need to deburr graphite electrodes either. When wire-cutting graphite, it will cut very slowly when compared with copper; although, ironically, wire-cut speeds will *improve* significantly in the denser grades of graphites.

COPPER

Copper cannot be as easily machined as graphite, although it is quite soft when compared to steels. That is part of the difficulty. It produces a soft, "gummy" type of chip that tends to weld onto the cutting edges, which can result in "tearing" of the metal and poor finishes. Tellurium has been added to aid in machining. Tellurium copper (called "Telco") will allow a "freer" flowing chip. Copper will always throw a burr on all the "off" edges of the cut. Deburring operations can be a significant cost factor, especially when specifications require dead-sharp edges or fine details that deburring operations would destroy.

A significant advantage to support the use of copper electrodes is the process called "discharge dressing," which was explained earlier. Copper readily discharge dresses with very few problems. The advantages of this method of fabrication are: unattended operations (CNC), dead-sharp edges with no burrs, and the ability to dress very fine or delicate details that conventional machining would destroy.

Copper electrodes cut by wire EDM are produced quickly, accurately and very efficiently, especially in the more complicated shapes and configurations. Wire-cut copper electrodes are an excellent example of wire-cut and vertical machines complementing each other in manufacturing. The electrodes can be manufactured "CNC accurate," unattended, and in quick enough turnaround to prove very cost-effective. This method is invaluable in complex shapes or very thin sections where heat or tool pressure would make this operation practically impossible. In addition, there are no burrs produced by using this method. Careful considerations in the placement of the "cut-off" tab left by the wire will leave little, if any, final finishing of the electrode.

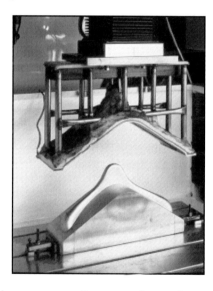

Figure 3.6   To reduce the weight of large electrodes, a "bird cage" frame supports a light weight "galvano process" electrode.

"Galvano process" copper electrodes are used when the large electrode size would make the electrode very expensive, in addition to being too heavy to allow adequate servo response of the EDM machine. An epoxy transfer similar to the type used by the abrading process will be made. This in turn is placed into an electrolytic bath, and copper is electrolytically deposited in thickness up to 0.200". Electric cables are then fitted to the inside of the copper shell and then it is partially filled with an epoxy liner for strength. It is then removed from the epoxy model to be prepared for mounting to the EDM machine.

Not to be overlooked is the "specialty" electrode market. Custom molded electrodes can be made to a customer's specifications from a part print or model. The electrode material is a 70/30 mixture of tungsten and copper powder. This is compression molded in a model of the cavity and sintered in a furnace to form the finished electrode. Electrodes fabricated in this way can maintain tolerances of ± 0.001".

Figure 3.7   View of the "business end" of galvano electrode and finished mold core.

This type of electrode is often used for multiple cavities or production EDMing of repetitive shapes, such as those commonly found in the turbine industry. The cutlery industry is also a large consumer of this type of electrode. Other good candidates for this type of electrode would be multiple electrodes that are complex or requiring fine detail such as logos, or large coining dies that would be labor intensive and not cost-effective if conventionally machined or fabricated.

GRAPHITE
Milling, turning, and grinding can all be done much faster in graphites, but caution must be used to prevent chipping at the edges or end of the cut. In milling and turning operations, cut "onto" an edge and slow down the feed rates when approaching the end of the cut. This will prevent the cutting forces from "pushing" the remaining unsupported material and fracturing it, usually chipping the edge. Generally speaking, the softer and less dense the graphite, the greater the possibility of chipping or "break-out" occurring.

As mentioned before, when using graphite, more than just machining it must be considered. The "chips" from graphite are a micro-fine black powder or dust that is not only a nuisance, but is highly abrasive to precision machine-ways, and can also prove a hazard to electrical equipment as well. Any shop that is going to make

Figure 3.8   An example of "wire-turning" a delicate copper electrode using a rotation device. (Courtesy of System 3R, USA, Inc.)

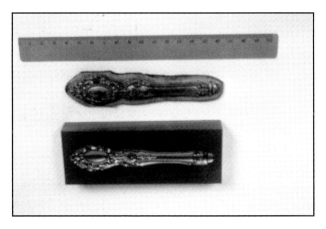

Figure 3.9   A "molded" copper tungsten electrode used in the cutlery industry. (Courtesy of Keltool, Inc.)

graphite electrodes more than just occasionally should have an efficient vacuum system installed with inlet nozzles at each work station.

For years, wire-cutting graphite electrodes has been considered slow; however, recent tests show cutting speeds increased significantly in the better grades of graphite — those with high density and small particle size. Test results recently conducted by an independent shop have determined that zinc- coated wires perform from 35% to 50% better than hard brass. These results show that a zinc-coated copper wire produced the fastest cutting speeds — 16 sq. in./hr. in an Angstrofine graphite, and almost 12 sq. in./hr. in a superfine graphite. Cutting the same materials with hard brass wire produced speeds of 11.5 sq. in./hr. and 6 sq. in./hr., respectively.

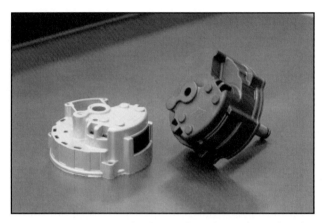

Figure 3.10   An example of a graphite electrode produced
by the abrading process. Diecast part is on the
right. (Courtesy of Sonex Division, ExtrudeHone.)

Certain applications using graphite lend themselves to "abrading." This is the
process in which a model or pattern of the electrode is made. From this model, an
epoxy transfer tool is made. This will be the actual abrading tool and it is charged
with a silicone carbide grit that will give it its cutting ability. It is then mounted in
a special machine above the graphite blank. In an oil bath, and under high pressure, it is forced into the block of graphite while orbiting, essentially "crush-forming" the graphite electrode. This orbit can vary from 0.020" to 0.200". This is quite
useful for large, complex electrodes used in high production or accurate redressing
operations such as electrodes for large forming dies, and automotive applications
such as crankshafts, transmission housings and castings with multi ribs and bosses
that would be tedious and time-consuming to make and redress.

Another method of preparing graphite electrodes is "ultrasonic" machining,
which is virtually a small-scale abrader. The tool that will make the electrode is
called a *sonotrode*. The master pattern is affixed to the sonotrode and the assembly

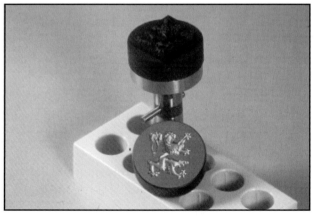

Figure 3.11   A graphite electrode produced by ultrasonic
machining. (Courtesy of System 3R, USA, Inc.)

is vibrated at about 20,000 Hz while circulating an abrasive slurry between it and the graphite blank. This rapidly erodes the shape of the electrode into the face of the graphite blank, resulting in fast, highly detailed (0.002r) electrodes that can be easily and accurately redressed. An example of electrodes made this way would be those used in EDMing fine coining or embossing dies.

RAW MATERIAL COSTS

So far, we have examined the actual physical properties of different electrode materials, and some of their better-suited applications. The last factor in the electrode selection process is the cost of raw material. The range of costs for raw material is as widespread as the number of types and grades of material. As a rule, the best grades of graphite tend to run approximately five times the cost of the least expensive. Figure 3.12 reflects only commonly ordered sizes, as this is the way most material is ordered — by size. Actually, the best way to calculate the cost of electrode material is by the cubic inch.

As you can see in Figure 3.12, there is a wide disparity in list prices of copper and copper tungsten. Obviously, the tungsten "sintered" product is much more expensive, but it can sometimes prove to be the better value by reducing the number of electrodes required when facing applications resulting in high wear, such as carbides. Also, you will see that copper tungsten is only available in 8-inch lengths,

MATERIAL	SIZE/SHAPE	APPROX. COST*
Round Graphite		
Infiltrated	1" Dia. × 12"	$104.00
Super Premium	1" Dia. × 12"	$ 72.00
Premium	1" Dia. × 12"	$ 59.00
Medium	1" Dia. × 12"	$ 30.00
Low	1" Dia. × 12"	$ 19.00
Round Copper		
Copper	1" Dia. × 12"	$ 23.00
Copper Tungsten	1" Dia. × 8"	$177.00
Bulk Plate Graphite		
Infiltrated	1" × 4" × 6"	$185.00
Super Premium	1" × 4" × 6"	$143.00
Premium	1" × 4" × 6"	$102.00
Medium	1" × 4" × 6"	$ 47.00
Low	1" × 4" × 6"	$ 25.00
Bulk Plate Copper		
Copper	1" × 4" × 12"	$ 95.00
Copper Tungsten	1" × 4" × 8"	$530.00

Figure 3.12   List prices of electrode materials, based on national averages as of August 1997.

so the chart does not reflect a true "apples and apples" comparison of copper and copper tungsten. The reason for this is because the initial demand for copper tungsten was for welding electrodes used in the auto and sheet-metal fabrication industry, before the prolific use of EDM. The standard lengths they specified were in 4- and 8-inch lengths. Since there are far more welding operations in America than EDM operations, the suppliers have not made any great efforts to change over to the standard 6- and 12-inch lengths in which most other materials are supplied.

The price comparison of raw materials alone can't direct a user in the best selection because, very seldom is the material used in its raw state or blank. The actual cost of an electrode is not based upon how much the raw materials are, or even material's cost plus the cost of fabrication. The actual cost is way down on the bottom line. Many questions have to be answered before the true cost of an electrode can be determined. Just like the rest of the EDM process, there are many variables affecting electrode selection that have to be carefully evaluated.

Specific questions are, for example:

"How much is the blank electrode material?"

"How long will it take to make?"

"How difficult will it be to provide good flushing?"

"How many electrodes will be needed?"

"How fast will it do the job?"

"How much wear will be encountered?"

"Will the cavity have the right finish?"

"Will I have to 'fight' the job?"

Only after evaluating the final cavity/cost per electrode can the true price of an electrode material be assessed.

SUMMARY

After learning about material properties, fabrication methods, wear and finishing characteristics, and raw material cost, we might think we have it all figured out. Not quite. Based upon the wear parameters outlined in Figure 3.5, we must decide whether to use graphite or copper, whether or not to make multiple electrodes, and, if so, how many? Is redressing necessary or cost-effective? What about flushing and other variables? The make and model of the EDM machine itself will also affect one's decision.

As with most EDM operations, a simple, easy way to do a job is not always possible; and, in my own experience, an "easy" job is usually the exception and not the rule. This book was not written to solve *all* of your EDM problems — but hopefully it *will* make your electrode material selection easier.

Chapter 4

FLUSHING

THE KEY TO SUCCESSFUL EDM OPERATIONS

There is an old adage, as old as EDM itself: "There are three rules for successful EDMing; flushing, flushing and flushing," and this is applicable in every case. Other than knowing the fundamentals of on-time/off-time and basic electrode material selection, the most important element of successful electrical discharge machining of *any* material is flushing. There are several different methods, but the concept is the same for any application.

Flushing is the process of introducing clean dielectric fluid into and through the spark gap. This serves several purposes:

1) introduces "fresh" dielectric to the cut;
2) flushes away the "chips" and debris from the spark gap; and
3) cools the electrode and workpiece.

As described in Chapter 2 on EDM theory, it is established that there must be off-time to allow the dielectric sufficient time to "recover." The combination of spark duration, dielectric "strength" and flushing efficiency are all factors determining the amount of time the power is off. Since *off* is exactly that - *off* - no work is being accomplished. We would like to keep this to a minimum; and the better the flushing conditions, the less off-time is required, and the efficiency of the entire operation is increased.

PLANNING

The most successful EDM jobs are the ones utilizing careful planning from the start. The concept of "designing around flushing" is the most prudent. From the blueprint stage, an experienced EDMer will study the cavity or detail and visually determine where and how the electrode or cavity is to be flushed. Conditions such as surface area, corners, "dead areas," side draft, turbulence, cavitation, gas evacuation, and secondary discharge are just a few. To the inexperienced EDMer, this may seem overwhelming, but all these things become second nature with time and experience.

Each style of flushing has its own properties and applications, and choosing the correct one is not always a straightforward proposition. Many times, the most logical choice is not practical, i.e., slots or other details that are too deep and/or too narrow to drill flushing holes in the electrode. So, just like a lot of other choices in EDM, a compromise must be made, usually opting for the next best method.

Many times, reasons such as, "it's too much trouble," or "there's not enough time," are given for not planning on, and providing for, the best flushing conditions economically possible. This attitude is almost always counterproductive. EDMing with optimum or near-optimum flushing conditions will always be *less trouble*, in most cases many times *faster*, and almost always *safer*. Not providing good flushing conditions where possible, in order to "save" money or time, is *false economy*. If you are an entry-level EDMer, learn this rule. If you are an experienced EDMer, don't forget it.

TYPES OF FLUSHING

There are three types of flushing. They are as follows.

1) **Pressure**. This the most common type of flushing, often referred to as injection flushing. The oil is forced through the spark gap, either through holes in the electrode or from holes in the workpiece itself.
2) **Suction**. The opposite of pressure flushing. The electrode and workpiece are prepared in the same manner as pressure flushing situations, but instead of the oil being "pushed" through the gap, it is "pulled" through by vacuum.
3) **Jet or Side**. The least efficient method, but far better than none at all, is jet flushing, or side wash. This is the strategic placement of hoses or flushing "wands" to direct the stream of oil to flush the gap during pulsed electrode movement.

PRESSURE FLUSHING

As mentioned above, this is the most prevalent method of flushing today, for several good reasons. All machines provide outlets for pressure flushing, but a few do not offer vacuum flushing. Also, pressure flushing (also called injection flushing) is much easier to learn and control. Gauging the flushing pressure can be done visually as well as by referencing pressure gauges.

Relying upon pressure gauges alone can be very misleading. There will be a large difference in pressure readings between a cavity that is "sealed off" during the burn and a side-flushing application where hoses with different sized nozzles and wands can require much more volume to flush properly.

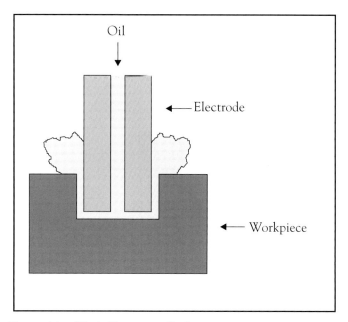

Figure 4.1   "Sealed" pressure flushing.

For general use and ease of operation, pressure flushing is the simplest and easiest method. The best conditions are when the detail allows for "sealed flushing." That is, when the electrode is in the "burn" position, all of the face or the edges of the electrode are engaged in the cavity. This forces the dielectric to flow across and through the spark gap (see Figure 4.1).

An example of this application would be through-hole flushing "in the solid," or finish burning with an electrode slightly larger than the roughed-out cavity. If the electrode has steps or is "staged," there are certain "tricks" one can use to prevent loss of pressure or vacuum (see "Tips and Tricks" later in this chapter for more information).

In pressure flushing applications, the oil is introduced through holes in the electrode itself or from holes in the material where the cavity will be. It is delivered to the electrode through hoses and flexible tubing, attached to the electrode itself, the tooling above it or, on some equipment, through the machine spindle. These same hoses can be attached to the workpiece itself, mounting fixtures, or through a "flushing pot."

FLUSHING POTS

Another common example of pressure flushing is represented in Figure 4.2. This is using a flushing "pot" or plenum chamber beneath the workpiece. This introduces oil into the work area through the workpiece instead of the electrode. This can be applied to many types of work, and is especially useful in moldmaking because of the predrilled ejector and core pin holes already existing within the cavity.

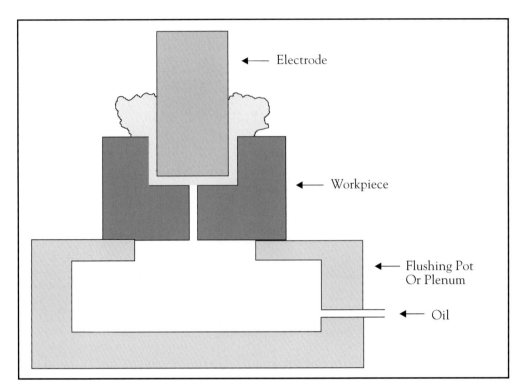

Figure 4.2   Pressure flushing using a flushing "pot."

Many times it is not practical or possible to drill the required flushing holes in the electrode. Very long or narrow electrodes are one example, and very hard electrode materials such as tungsten or copper tungsten are another. Using the preexisting holes in a cavity, whether "pushing" or "pulling," will save the time and trouble of drilling long, small-diameter holes in the electrode.

SECONDARY DISCHARGE

At this point we have to leave the discussion about flushing to discuss a little EDM theory and applications. This is necessary if one is to understand not the *differences* between pressure and vacuum flushing, but the *reasons for choosing* one or the other.

When using pressure flushing, the electrode tends to run cooler and the workpiece will be warmer after a period of time, especially during roughing (high-amperage, low off-time) operations. This is because the cool, clean dielectric flowing through the electrode acts as a radiator or heat sink, carrying away heat from the body of the electrode that has been transferred by convection from the area of the spark gap.

In turn, the oil is heated rapidly as it travels across and through the spark gap, carrying heat and contaminants along with it. This very hot oil tends to heat up the side walls of the cavity, and the surrounding material tends to expand slightly, closing in around the sides of the electrode. This slight "closing" effect, in addition to the still conductive particles passing by the finished cavity walls, can result in what is called "secondary discharge." This is the naturally occurring principle of an electric potential seeking the path of least resistance. (Doesn't lightning usually strike taller [closer] objects?) The combination of thermal expansion of the side walls, and sparks striking chips that are grounded to the finished cavity walls, can result in tapered side walls and poor finishes. As long as this tapering is moderate and will clean up during finishing operations, this should not present a problem; but if the side walls must be very straight, accurate, and display a uniform finish from top to bottom, then vacuum flushing could be the solution.

SUCTION FLUSHING

This method is used when the accuracy and straightness of cavity side walls are imperative. The primary reason for using vacuum or suction flushing is to eliminate the possibility of secondary discharge on a finished side wall. Clean, cool oil is drawn past the finished side walls, preventing secondary discharge (see Figure 4.3). While this is a common problem with manual machines (cutting only in the Z- axis), CNC machines have the luxury of cutting in any axis or direction with the entire side of the electrode, and as long as sufficient stock is left for finishing, CNC machines seldom have to employ vacuum flushing.

While vacuum flushing does offer accurate machining capabilities, there are a few drawbacks. With vacuum or suction flushing, the operator loses the visual as-

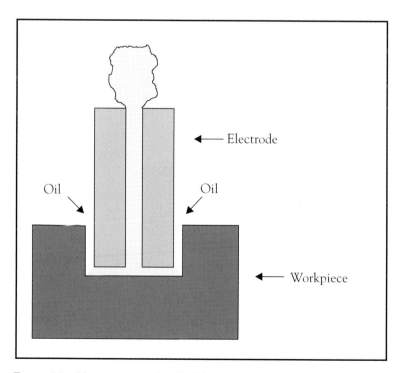

Figure 4.3   Vacuum or suction flushing.

pect of gauging flushing conditions as there is nothing for an operator to see other than the vacuum gauge itself. With pressure flushing, there is a visible stream of oil, and the operator can gauge the volume and velocity of the oil by observation, in addition to gauge indications. Even inexperienced operators can quickly develop this "sense," for regulating flushing pressure, much like hearing the sound of the burn itself.

Suction flushing does not provide the ability to see what kind of "setting" you have, and true gauge readings are not stable until the cavity is "sealed off" and the burn is in progress. Even then, pressure can continue to drop as the spark gap becomes smaller and is contaminated with chips and debris. If this pressure is allowed to drop too low, the electrode can be pulled from the mounting fixture, or, in severe instances, the vacuum conditions can pull the workpiece from a vise or magnetic chuck!

A word about flushing pressures: we are dealing with a light oil, with properties not unlike a hydraulic oil. This is my point — hydraulics — which is a very strong and undeniable force to be reckoned with, whether it is being "pushed" or "pulled." If you have any doubts about the power of hydraulics, consider the fact that the one-inch diameter cylinder in a hydraulic jack is enough to lift an automobile! If the surface area of your cavity is several square inches, there is an incredible potential for a hydraulic situation to develop.

If pressures are too high, excessive corner wear, slow machining times, and inaccurate cavity bottoms and side walls can occur. If this pressure is high enough, the servo system will not be able to overcome this resistance, and the electrode cannot advance into the "discharge area" of the spark gap. No discharge means no

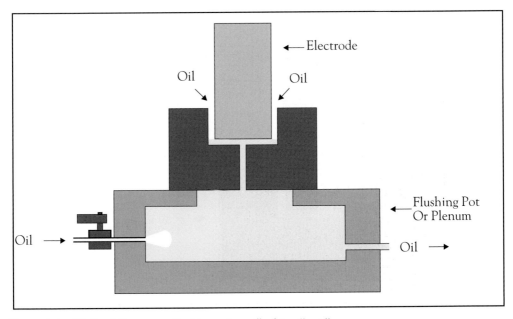

Figure 4.4   "Fail-safe" vacuum flushing using a flushing "pot."

work. Likewise, the opposite effects, including shorting, can occur with suction flushing if pressure levels drop too low.

Accurate gauges, responsive servo systems, and good EDM "sense" will go a long way in making vacuum flushing applications successful. Providing "fail-safe" methods will also help. Placing the entire setup on a flushing "pot" or plenum can provide the needed venting through a relief valve which is required to prevent the vacuum from reaching a critical level (see Figure 4.4). Drilling a small diameter vent hole in the electrode, or by regulating an additional relief hose via a manifold, is another possibility. The amount of vacuum and venting will have to be determined on a case-by-case basis. Just like most of the "finesse" operations that are required in EDM, this also is a "cut-and-try" method.

JET OR SIDE FLUSHING

This is the least efficient (and our last) choice, but many times due to size or shape restrictions, this is our only choice. This method almost always has to be augmented with a timed pulse or jump of the advancing axis (usually the Z-axis). This is actually a "mechanical assist" to the flushing procedure. The withdrawing electrode causes fresh, cool dielectric to be drawn into the cavity area, "diluting" the contaminated oil. The return stroke of the advancing electrode displaces the oil in the cavity and forces out most of the contaminated oil. Without this jump or pulse, it would be almost impossible for the clean oil from the flushing wand(s) to flow through the small spark gap.

As you can see in diagram 4.5, the flow of dielectric is introduced from the side. As it enters the first side, it is relatively clean and cool. As it passes through the bottom and actual working area, it assimilates the bulk of the heat and contaminants. Even with the help of the pistoning electrode, this is very ineffi-

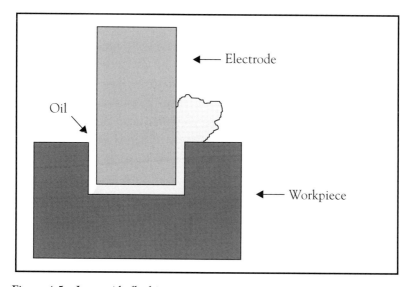

Figure 4.5   Jet or side flushing.

cient, many times trapping particles in corners or other "dead areas," resulting in dc arcing and pitting. Finally, all of this "mess" has to defy gravity and flow up and out of the opposite side, and past the finished side wall. This can be assisted with a suction hose placed at the exit side to "vacuum" the debris out. Even with a vacuum hose on the exit side, pitting of the finished vertical wall can result due to secondary discharge.

Obviously this should be our last choice, but it can still work relatively well if attention is paid to flushing nozzle pressure and positioning (especially with multiple outlets).

FLUSHING PRESSURE

There are many different machines made by many different manufacturers from many different countries. Some use English graduated dials and gauges and others use metric. When relying on gauges for pressure or vacuum settings, converting pressure settings from one machine to another is sometimes necessary. The formula to convert pounds per square inch to kilograms per square centimeter is:

$$1 \text{ lb./in.}^2 = 0.0703 \text{ kg./cm.}^2.$$

Rather than recommending a specific pressure setting (all applications present different sizes and locations of holes, surface area, etc.), go on the theory of "volume not velocity." It is far better to have a lot of low pressure holes in the electrode or cavity than one or two "jets" at high pressure. This can deflect the electrode, increase corner wear, cause deformities in the cavity's surface, and make the servos work harder than they should.

FLUSHING - SPECIFIC APPLICATIONS

TUBING A HOLE

Probably the least troublesome of all applications is "tubing" a hole. This is done with brass, copper, or tungsten tubing, and used most of the time with pressure flushing. Only with some "exotics," or very small diameters (0.020" or less), will one encounter difficulty EDMing a hole. Electrode rotation is not absolutely necessary, but it is a definite plus for this application because it aids in flushing and affords uniform electrode wear. The "footprint" of this burn has a small cross section, and the dielectric does not have to flow very far before escaping the spark gap.

The larger the "swept area" or working surface of the electrode, the more difficult the operation will be unless more flushing holes are added. Unfortunately, there are no set rules for the number of flushing holes per square inch. The number is arbitrary, with the thought that it is better to have too many holes than too few. Keep in mind, however, the larger the total area of flushing holes, the less actual pressure should be used.

The placement of these holes is just as important. Sharp corners, deep cavities, and particle trapping are all serious considerations when determining the placement of flushing holes.

In the following example, flushing oil is introduced through 1/8" diameter silicon tubing (available from any aquarium supply store). The main "feeder" hole is drilled slightly undersized with a 0.120" diameter drill. This will provide a sealed, slip fit for the tubing.

The problem with this electrode configuration will be the loss of flushing pressure through the open hole that is not yet engaged or "sealed off." Hydraulics obey a very basic law of physics and will take the path of least resistance; in this case, the open hole is far less resistant than traveling through the small spark gap, therefore it must be blocked off until it is needed.

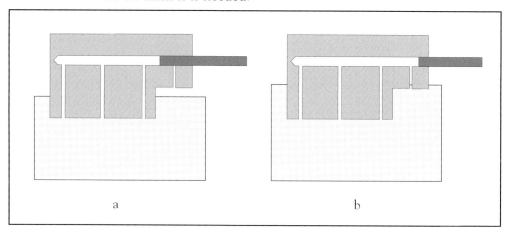

Figure 4.6   "Tubing" a hole for flushing.

To prevent pressure loss through the exposed, open flushing hole, slide the tubing past this hole, effectively sealing it off (Figure 4.6a). Set the micrometer stop or CNC control to stop the downfeed as the upper step comes in contact with the top of the workpiece. At this point, slide the tubing out far enough to allow flushing through this hole (Figure 4.6b). Adjust flushing pressure and resume machining.

STEPPED CAVITY REENTRY

Another difficulty often encountered is reentering a stepped cavity with a finishing electrode. As you can see in Figure 4.7, the open part of the detail allows the oil to escape and we do not have sealed flushing. This will result in poor cutting speed, excessive Z-axis servo motion, and possibly dc arcing.

By clamping or gluing a thin strip of shim stock over the open area (see Figures 4.8a and 4.8b), this will seal off the open part of the cavity and prevent the oil from escaping, and will significantly improve cutting speed and efficiency. When the stepped portion of the electrode approaches the shim, it can either be

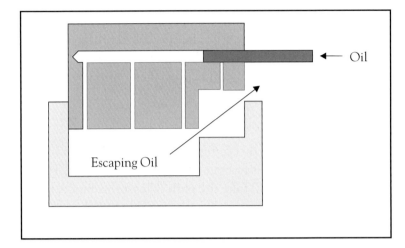

Figure 4.7   Flushing oil can escape from a stepped cavity.

removed or allowed to burn through the shim itself.

Lead tape has also been used with success, but make sure of its conductivity before starting. The adhesive may insulate the metal tape from grounding, so a simple strap clamp on the edge of the tape should prove sufficient.

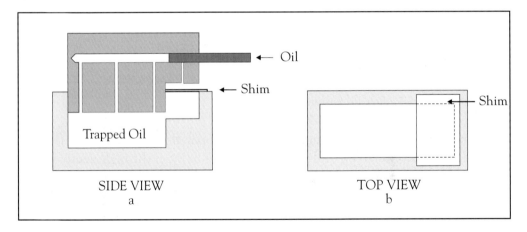

Figure 4.8   Using a shim to trap oil in a stepped cavity.

DESIGNING FOR FLUSHING

The advantages realized by good initial planning and "designing around flushing" are: increased cutting speeds, improved finishes, and the reduction of arcing. A few disadvantages of providing optimum flushing conditions would be: the time required in the planning of and providing for the best flushing possible, and the free-standing "studs" or "spikes" that are left by the holes themselves.

As mentioned previously, planning the job, especially in the areas of electrode design and flushing, is all important in the ultimate success of any EDM operation. But even with the right attitude toward planning the job, there is a point where the cost of time and/or the tooling required to provide the "ideal" conditions become cost-prohibitive. As in most EDM operations, there are no set rules for any situation, and each must be evaluated on a "case-by-case" basis. The cost of the electrode material, electrode fabrication costs, tooling, set-up time, and operator and machine capability must be weighed against the budget allowed for a given operation.

But EDM presents an additional variable not usually factored into other, "more conventional" machining methods — the *safety* or *risk* factor. We all know Murphy's Law. Well, Murphy was an EDM operator!! If something can go wrong, it will, and one must consider these "chances" when quoting or preparing for an EDM job. Many EDM jobs are almost finished (and therefore expensive), when they arrive at the EDM machine. In such a case, additional allowances for planning or tooling expenses can be easier to justify. Remember the saying "Penny wise and pound foolish?" Don't make this same mistake. Taking the few extra minutes required to add proper flushing to an electrode can save hours of machine time, and perhaps make the difference between profit and loss.

As for the remaining material left by the flushing holes, the diameter of these "spikes" depends upon the actual size of the flushing hole itself and the amount of overcut. CNC machines have a definite advantage over manual machines because an orbit or translation of the table larger than the hole diameter will effectively machine away the spikes left by static cutting (see Figure 4.9).

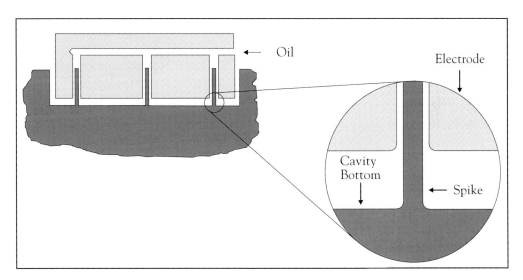

Figure 4.9 "Spike" diameter is dependant on the size of the flushing hole and the amount of overcut.

By machining the cavity "in the hard," as are most moldmaking operations, the free-standing spikes left by the flushing holes can be easily snapped off with needle-nosed pliers (be careful not to damage the cavity walls or impale yourself on the brittle shards of the spike!). If the cavity is soft or the material is ductile, bend the spike back and forth, and fatigue and work hardening will cause it to finally separate.

In operations where static machining is the only option, taking a little more time in preparing the electrodes before their use will pay off substantially in the long run. By drilling flushing holes at an angle, to intersect larger "feeder" holes that are drilled from the back of the electrode, the spike will be machined away as the ram descends, leaving a much smaller "pimple" on the bottom of the cavity instead of the tall spike (see Figure 4.10). Plug the cross-drilled holes or use a backing plate to prevent oil from escaping from these holes. (*Note*: Pay careful attention to the flow patterns and cavity shapes before drilling any holes. The chance of creating "dead areas" is increased by drilling angular holes which can direct oil away from a critical area.)

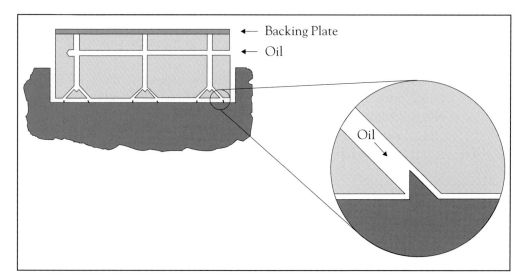

Figure 4.10   Angled flushing holes leave a much smaller "spike."

These flushing hole "witnesses" can be removed entirely with a symmetrical cavity or detail (a mirror image from center) by drilling flushing holes in an offset pattern instead of symmetrically from center as is often done.

In the example shown in Figure 4.11a, the flushing holes are drilled 0.100" off center. This is not enough to detract from good flushing conditions, and more than enough to ensure total clean-up later.

After the desired depth is obtained, withdraw the electrode, and with needle-nosed pliers, snap off the free-standing spikes as close to the cavity bottom as you can. (This is much easier if the cavity detail is prehardened). Rotate the electrode 180°, reduce power, and re-enter the cavity (Figure 4.11b). Cutting will be erratic until the electrode "settles in," but this time is insignificant when compared to the time required to grind and polish away these marks.

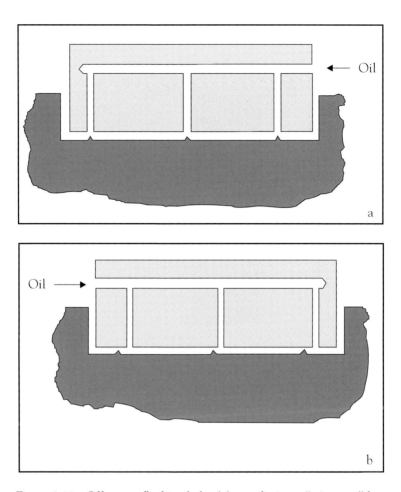

Figure 4.11   Off center flushing holes (a) can eliminate "witnesses" by rotating the electrode (b).

DEEP RIBS AND SLOTS

Another common challenge to the EDM operator is the narrow, deep rib or slot, which is often so narrow or deep that it is practically impossible to drill the small-diameter flushing holes that will be required. For this example, the electrode can be prepared as follows. Grind the electrode to thickness, then select the same diameter copper tubing (allowing, of course for overcut). Grind the width of the electrode one diameter undersized, but with the matching radius running the length of each side (see Figure 4.12). Carefully nest and glue the copper tubing into this radius and clamp until set (see the section below concerning glue). The manifold is easy to make and will have outlet holes on the correct centers.

This works very well within tolerances of ± 0.005". If greater accuracy is required, fabricate a second electrode of solid copper or graphite to reenter and finish the cavity. Even without good flushing, the small amount of stock remaining should not prove too difficult to remove.

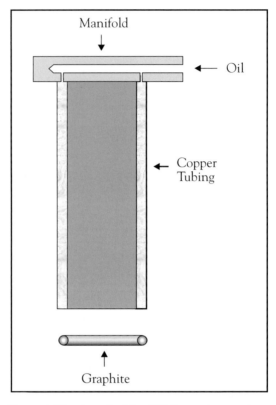

Figure 4.12   A method for EDMing a deep cavity.

Another method to aid an operator to successfully EDM a narrow, deep cavity is to make small manifolds or diverters. These are most easily constructed out of scrap pieces of graphite or old and unserviceable graphite electrodes. Use your imagination on these as to the flow characteristics and placement. These can help direct the flow of oil in the desired direction without having to readjust flexible flushing wands to get the flow "just right." They can be glued in place or held in place with the thumbscrews, vise, or toolholder. If they are not glued to the electrode, they can be used indefinitely.

An example is shown in Figure 4.13. In example "a," you can see how the opening of the device directs the flow of oil along the face of the electrode. The electrode in "b" has been partially relieved on one side to channel the oil all the way down to the bottom of the detail. This will improve machining speed and, more importantly, safety. However, this relief will allow material on this side and will have to be removed by rotating the electrode 180° and reentering. Slots and grooves can be used to channel fresh dielectric into a cavity when providing flushing holes would prove too difficult.

To flush round electrodes that are too long or too small to drill holes through their length, grind or file a flat along the length in one or more places, as in Figure 4.14a. This flat doesn't have to be pretty or accurate, as long as oil can be directed down the length of the electrode to help flush the tip. Obviously, this method will

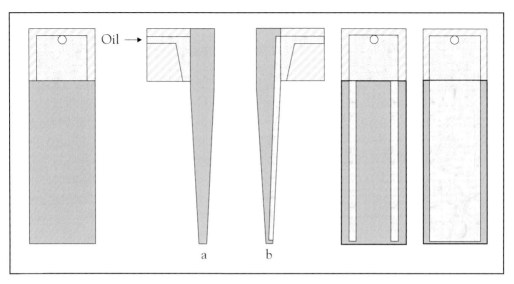

Figure 4.13   Using "diverters" to EDM a narrow, deep cavity.

require electrode rotation, otherwise these same flats or grooves will be left in the cavity walls.

A similar method, but providing much better flushing, is grinding or milling a slot down the electrode's length (Figure 4.14b). Do not, however, cut to centerline or a spike will be left standing in the cavity bottom. This electrode also will have to be rotated. Flushing wands directing oil down into the slot as it rotates will speed up this operation considerably.

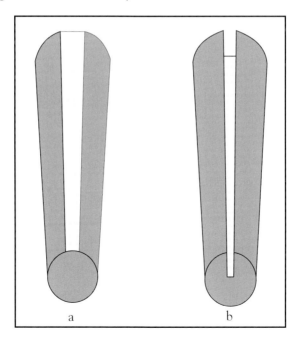

Figure 4.14   A slot cut down the electrode's length will improve flushing.

GLUE

Throughout this section, the recommendation to "glue" or "secure with glue" has been made several times. This means just what it says, but with one qualification: whatever glue is used *must* be conductive. To be strong and resistant to heat and oil goes without saying, but it must also be able to conduct electricity.

Many glues are available expressly for this purpose and are available through most EDM supply houses. Other glue "kits" are available and require adding nickel or silver compounds to the glue to provide conductivity.

What has worked very well for me is to use a smooth mill or second-cut file to scrape a small amount of graphite into a fine powder. (*Note*: Copper infiltrated graphite will offer higher conductivity.) Mix this powder with ordinary "super-glue" to form a thin paste (this will also make the glue easier to work with). Using a toothpick or needle, carefully apply this mixture to your electrode/tubing/holder/ shank and assemble. Before using, check entire assembly for electrical continuity. Failure to do this can result in a crushed and shattered electrode and possible damage to the machine.

SUMMARY

Vertical EDM presents many adversities, flushing being one of the most difficult to overcome. Careful planning before the part or electrode is made will often prove the single greatest factor of smooth, efficient EDM operations, but we don't always have that luxury. Most often we are forced into "making something work."

Many times, flushing will be the most important part of an EDM operation, but in just as many cases, flushing can be the most difficult to administer. Don't cheat yourself or your shop's operation by the natural tendency to skimp on the time necessary to provide adequate flushing. As stated earlier, this is almost always "false economy."

There are very few "textbook" examples in real-world EDM, so you can't always rely on "textbook" answers to see you through. Use existing and proven methods on the "easy" ones, and a combination of your imagination, common sense, and experience on the rest.

Chapter 5

ORBITING

INTRODUCTION

Since the advent of CNC EDM, not much attention has been paid to orbiters or orbiting. Often referred to as "the poor man's CNC," it does have several advantages to offer, but is still a long way from a CNC. First, we will examine the reasons why we would elect to use orbiting, then we'll take a look at the different types and styles of orbiters available, and finally, the features and capabilities each type can offer.

TRADITIONAL MANUAL EDM METHOD

First, we should examine a routine procedure using a manual machine. This will set up the parameters we will use later for comparisons against EDMing while orbiting.

Since we are dealing with manual EDM, we are required to make multiple electrodes — roughers and finishers. All electrodes must be fabricated undersized to allow for the spark gap or overcut, relying heavily on the technology supplied by the machine's manufacturer. Charts within the manuals of the machine should provide overcut information based upon electrode material, workpiece material, amperage, and the finish desired. With this information, the spark gap or overcut can be determined. The operator, through simple subtraction, can arrive at the actual sizes of both roughing and finishing electrodes.

Most experienced operators have memorized their "pet" machine settings and overcut allowances and that's good. For the less experienced users, we will try to demonstrate how we arrived at these settings. For discussion's sake, we will use arbitrary numbers that are easy to use, but not necessarily correct.

Our hypothetical cavity is 1-inch square by 1-inch deep, with a tolerance of ± 0.005" of all dimensions. Consulting the machine's cutting conditions manual says the setting for the finish we want will have an overcut of 0.005" per side. Make this electrode 0.990". The EDM manual says the roughing electrode at the maximum amperage allowed for this surface area will have 0.015" overcut per side. Make this electrode at least 0.030" smaller than the finisher. (Personally, I have usually allowed an extra 20% per side. Side-wall expansion and secondary discharge can quickly encroach upon this safety margin.) Using these numbers the roughing electrode should be 0.036" smaller than the finisher, making it 0.964".

Theory: If we are to err, let it be on the side of safety.

Practice: We can always take *more* material *out*, but it's pretty tough to *put it back in!*

By using elevated on-times to protect the electrode, the rougher should suffer very little wear and should leave at least 0.005" of material remaining on all surfaces for finishing.

Using 65 amps per square inch as the maximum allowable current, we can determine the power setting for finishing in the example above:

0.005" (per side) × 4 (1" sides) = 0.02 (sq. in.) × 65 (amps) = 1.3 amps.

Let's assume that when using this power setting at the frequency necessary to obtain the specified finish, the electrode will incur 20% wear. What makes this condition so bad is that we must use this very low "wear setting" to cut the entire Z-depth all over again! This means that at 20% wear, only 0.800" of our 1" deep cavity will be to size. The cavity bottom will be to depth within 0.001" since the electrode only had to remove 0.005", but the side walls will be badly tapered.

So it seems that our finisher does not really finish our cavity. We must stop and redress the worn electrode or replace it with another one. Then we must reenter the cavity to remove the remaining 0.200" of material. Still incurring 20% wear, we will be leaving as much as 0.040" in the corners of the next burn. We must, again, redress or replace the electrode and continue. This finisher will leave 0.008". One more electrode and we finally finish the cavity within 0.0016" of nominal (see Figure 5.1).

ROUGHER	0.995	DEPTH OF CUT
1st Finisher	1.000	Depth of Cut
	- 20%	Wear
2nd Finisher	0.200	Material Remaining
	- 20%	Wear
3rd Finisher	0.040	Material Remaining
	- 20%	Wear
4th Finisher	0.008	Material Remaining
	- 20%	Wear
5th Finisher	0.0016	Material Remaining

Figure 5.1   Traditional steps for EDMing a cavity.

Summary: The cavity will meet blueprint finish and dimensions after using one rougher and five finishers.

This would require either one rougher and four finishers (and the time and material to make them), or one rougher and one finisher plus the time and support equipment required to redress four times.

Now let's look at an alternative.

MANUAL EDM USING ORBITING

We will attempt the same cavity again using the same machine and the same roughing settings, but the rest of this procedure will be significantly different and substantially faster through the use of an orbiting device. So different, that instead of changing to the larger, finishing electrode, we will continue machining with the original rougher.

By using the same high on-times or "no-wear" settings of our previous attempt, our roughing electrode will have very little wear, and our cavity walls should have approximately 0.005" of material per side remaining. With an orbiter we can move the electrode *sideways*, engaging much more surface area. This means we can remove the remaining stock by moving 0.018" per side, (the difference in roughing and finishing spark gaps and the remaining 0.005" of material per side). This removes the remaining 0.005" from the side wall *all at once* instead of machining the Z-axis all over again! (See Figure 5.2.)

Summary: The cavity will meet blueprint finish and dimensions using only one or possibly two electrodes of the same size.

ROUGHER*	0.995	DEPTH OF CUT
First Finisher*	0.005	Depth of Cut
	- 20%	Wear
Second Finisher	0.001	Material Remaining
	- 20%	Wear
	0.0002	Material Remaining
*same electrode		

Figure 5.2   Using orbiting to EDM the same cavity as in
Figure 5.1.

Raw material costs and labor expenditures are less because of the reduction in the number of electrodes required per cavity and proportionately, the time and equipment required to produce them.

Cutting speed will increase because by presenting the entire side of the electrode to the cavity wall we can increase the current. This will not reduce the quality of the finish because the higher current is spread out over the larger surface area.

Surface finishes will improve because the electrode is moving across the material's surface, effectively "averaging out" the imperfections or flaws in the surface of the electrode. This would normally be "imprinted" into the surface by a static electrode.

Machining safety is increased because the electrode is moving around in the cavity, mechanically aiding the exchange of dielectric and providing more room for chips and debris to escape from the "open" or "away" side.

Figure 5.3 below shows several patterns and pattern combinations that could be used to machine a square cavity.

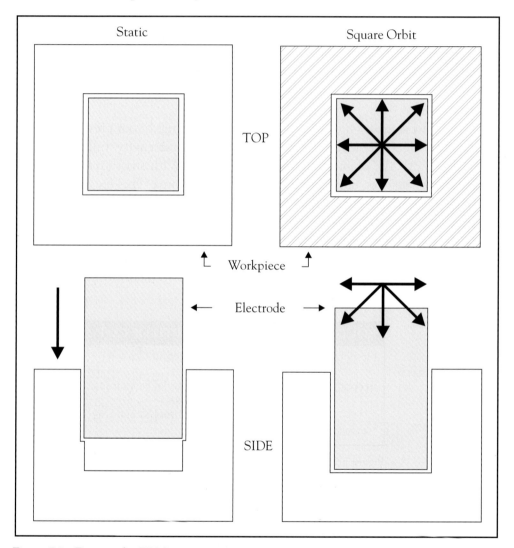

Figure 5.3   Patterns for EDMing a square cavity.

As you can see in this figure, static machining will produce excessive wear and long machining cycles because of the severely limited surface area presented to the electrode. On the other hand, orbiters can offer many different combinations of patterns to finish a cavity. By feeding the electrode on a 45° angle out and down simultaneously, orbiting can provide up to 3 square inches of surface area for much more efficient and safer machining.

COMPARISON

Figure 5.4 shows the differences in the requirements and results of both methods. Without question, orbit-assisted machining of this detail is much more efficient.

STATIC	ORBITING
requires at least five electrodes	requires only 1 or 2 electrodes
excessive "leading-edge" corner wear	uniform wear of entire surface
possible tapered sidewalls	dead-straight side walls
slow machining times	significantly faster
labor intensive	minimal operator involvement
sealed flushing mandatory	improved flushing due to motion
acceptable finish	an improved, uniform finish

Figure 5.4   A comparison of static and orbit-assisted machining.

SUMMARY ANALYSIS

Machining "sideways" through orbiting allows the following.

1) **Reduced machining time**.
   a) Presenting a much larger surface area of the electrode to the cavity allows the use of higher currents and frequencies without sacrificing finish because the additional power is spread out over at least 1 square inch. This will result in a substantially faster metal removal rate (MRR).
2) **Reduced, uniform electrode wear**.
   a) Machining with all sides of the electrode ensures uniform electrode wear instead of requiring only the leading edges of the electrode to do all of the work.
   b) Machining tangent to the vertical walls in the X- and Y-axes allows the removal of 0.005" × 1" instead of 1" × .005", increasing the work-to-wear ratio.
3) **Reduction in the number of electrodes required**.
   a) Costs are reduced by decreased consumption of raw materials.
   b) Further savings are realized from less direct labor required to make or redress multiple electrodes.
4) **Improved surface finishes**.
   a) Improved surface finishes due to electrode motion. This prevents the electrode from "imprinting" its EDMed texture into the workpiece.

b) By machining into corners at a 45° angle, adjacent walls of rectangular cavities will have consistent and uniform finishes despite any differences in surface area. This includes the cavity bottom, also.

5) **Improved accuracy.**
   a) Orbiting provides the ability to precisely control cavity sizes *mechanically* instead of relying upon electrode size and overcut.

6) **Reduced scrap.**
   a) Fewer rejects due to tapered side-walls, pitting or dc arcing. This is attributed to electrode motion which aids in flushing the debris from the cavity.
   b) The increased control over accuracy ensures consistent, predictable results.

7) **Reduced operator involvement.**
   a) Reduction of direct labor costs and support equipment involvement because fewer electrodes are required and proportionately less electrode fabrication.
   b) Polishing or benching of the workpiece is reduced or eliminated due to the improved surface finishes and removal of flushing hole "spikes" by orbiting.

Conclusion: By using an orbiting device, almost any detail can be executed faster, cheaper, more accurately, and with greater machining safety.

ORBITING SHAPES

There are many different shapes used in orbiting. By definition, the word orbit is from the Latin word *orbis*, meaning circle. Fortunately for us, orbiting by our definition is not limited to circles. The most typical shapes available, also called *loran*, are shown in Figure 5.5.

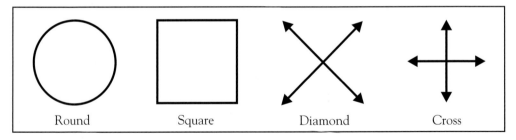

| Round | Square | Diamond | Cross |

Figure 5.5   Typical loran shapes and patterns.

There are other loran shapes and patterns available depending upon the make and model of the orbiting device, but the four shown should be sufficient in describing some of the advantages that orbiting offers. They can be used individually, in conjunction with others, and with dissimilar electrode shapes. They are described as follows.

1) **Round**. A circular pattern is used to enlarge round details or to incorporate corner radii on other geometric shapes. Using a square electrode with a 0.010" circular orbit will enlarge the cavity 0.010" per side while "wiping" a 0.010" radius on all outside corners. Using the orbiter in this method eliminates the necessity to incorporate the 0.010" radius on the electrode itself.

2) **Square**. A square pattern is used to symmetrically enlarge a square or rectangular cavity. This pattern is peripheral, that is, the electrode is fed outward the specified amount and then travels in a square pattern around the centerline. This helps "polish" the cavity sidewalls, but if this motion is used in only one direction, corner wear will develop on the leading edges. Also, if a particle is trapped and "rolled" between the electrode and the cavity walls, it can make linear scratches on the electrode that can be transferred to the workpiece.

3) **Diamond**. Also called, "tipped square." This is a very good pattern used to enlarge square or rectangular shapes. The electrode is fed out at a 45° angle and returns to center before feeding out again in another direction. This is very good for flushing and eliminates the chance of "rolling" a particle around the sides of the cavity. This pattern is especially useful when machining ribs or a detail that has substantially different surface areas adjacent to each other.

4) **Cross**. This is often mistaken for a square pattern. Like the diamond pattern, the electrode is fed out and returns to center.

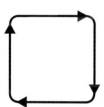

Figure 5.6   EDMing a cavity with a square electrode using a circular orbit produces this effect.

After selecting the orbit shape, some models can be fed out peripherally or in single or multiple quadrants, while others may be fed out in much finer increments down to 5°. The diagram in Figure 5.6 shows the effect of EDMing a cavity with a square electrode using a circular orbit. This eliminates the need to dress corner radii on the electrode. They can also be used in combination with other shapes as shown in Figure 5.7.

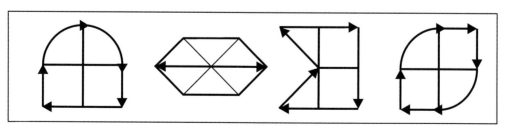

Figure 5.7   Many different patterns are possible by combining shapes.

TYPES, FEATURES AND CAPABILITIES

There are several styles of orbiting, depending on whether the orbiting device is mechanical or servo-driven.

Mechanical

The "cam-drive" types are the least expensive mechanical orbiting devices and are usually limited to a mode called "free-loran," but they are still much better than EDMing with a static electrode. Free-loran EDMing is orbiting the entire amount of offset movement while the machining rate is controlled by the Z-axis.

These types of orbiters are effective but not very "smart." That is, some of them will orbit the amount the operator sets, but they cannot detect if they are fed out too far or too fast. They are "deaf and dumb" insofar as the machine's servo system is concerned, and so they operate totally separate from the machine. If there is a flushing or clearance problem, the servo-system of the EDM machine will withdraw the electrode from the cavity; and when the conditions improve, it will then reenter the cavity and resume machining. This is seldom a problem when roughing "in-the-solid" as flushing is greatly improved by this orbiting motion. Finish machining is improved through improved flushing and better surface finish, but electrode wear is still a problem because of the fixed orbit. It will orbit to its full setting outward while depending upon the Z-axis servos of the machine to control machining, essentially "boring" the cavity. This presents the same problem of electrode wear as static machining — all of the work is done with the leading edges of the electrode. While flushing and finish are improved, machining rates will be slow and with considerable corner wear.

Other units are "semi-smart," meaning they are "gap-sensitive" to the extent that when they encounter a difficulty, they can stop or change what they are doing but they are not actually "talking" to the machine's control. Some of these operate by the following means.

1) When encroaching upon a preset spark gap (servo voltage), the orbiter speed will decrease or stop entirely. This prevents the orbiting electrode from being forced into a cavity wall, and allows the Z-axis time to withdraw until conditions improve enough to proceed.

2) In conjunction with decreasing motor speeds, some have additional features: Some can "collapse" or decrease the preset orbit, thereby increasing the spark gap to a level at or above the level that has been preset. This will allow machining to recover faster and prevent damage to the electrode or workpiece.

3) A pulse-timer can be set, much like the up/down pulse device on most vertical EDM machines. This forces the electrode to orbit out for a preset amount of time, and withdraw to center for another timed interval.

4) An insulated actuator rod rests on the bottom of the cavity. As long as there is downward pressure to keep this rod preloaded, it will allow a clutch device

to continue to orbit the electrode. When the pressure is released, the clutch will slip, allowing the orbit to decrease. If the Z-axis is withdrawn from the cavity bottom, this orbit will continue to shrink as the Z-axis rises. If the Z-axis is pulsed up and down to aid flushing, the actuator will force the orbit in or out accordingly.

SERVO-CONTROLLED OR CNC UNITS

The ability to machine with a servo-controlled orbiting axis provides much better control and several different methods. Servo-controlled means the electrode is "smart." That is, it won't allow itself to crash or run into a cavity wall because the servos will require it to maintain the correct spark gap. Some of these can be set up to obey their own control. This is done by allowing it to send a false signal to the machine control, telling it everything is all right, while the actual message from the spark gap is sent to the orbiter's control, allowing it to advance or withdraw the orbit or Z-axis.

Different styles include the following:

1) **Free-loran**. Like the simpler mechanical units, this style allows the loran movement while EDMing in the Z-axis. This is best reserved for rough machining "in-the-solid" as an aid to flushing.

2) **Lock-loran**. This is the process where the Z-axis is actually locked at a specific depth and the orbiter is gradually seeking the offset amount under servo control. Locking the Z-axis prevents it from withdrawing if difficulty is encountered. Since the orbiting device is servo controlled, it can reverse its pattern of movement to allow the debris or flushing problem to clear itself and then resume machining.

   This method is much better because the entire side of the electrode is presented to the cavity wall as described earlier, providing faster machine times, better surface finishes, and less electrode wear. Lock-loran (or "Z-lock") is useful for machining undercuts, t-slots, thread "rolling," and clearance tapers for stamping and extrusion dies.

3) **Step-loran**. The electrode is fed outwards in steps, usually radially. This can be done in conjunction with different depths of machining to leave steps in the part or to mimic CNC machining by allowing lower power settings to be used as the steps approach finished size.

4) **Vectoring**. The electrode is fed in a single, arbitrary direction. It is used to lengthen or enlarge a section of a cavity or detail by machining out to a predetermined dimension and returning to center.

(Some of the above terms may differ from vendor to vendor, but they are all basically the same.)

SELECTION

Hopefully, the items covered in this chapter will help an EDM shop to better understand the merits and benefits of applying motion to the electrode while machining. Shops that already have CNC EDM machines or orbiter-equipped manual machines are already quite aware of the advantages of electrode motion.

The logical followup to this topic would be "which one is right for me?" Obviously, every shop has different machines, jobs, methods, and budgets; so thankfully, I don't have to decide for you (after all, it's very easy for *me* to spend *your* money!). But I can offer some guidelines to use when it's time for you to consider one:

1) Carefully weigh your application(s) against the orbiter's features. While almost anything related to EDM and tooling seems to be expensive, don't limit your present and (more importantly) your *future* capabilities by buying short of your needs. Most of the time, when something is purchased to "just get by," that's exactly what will happen.

2) Make sure the weight of the prospective orbiting device is carefully matched to your equipment. Electrode weight is a very serious consideration, and for all practical purposes, *the orbiter is considered part of the electrode*. Excessive weight can be a serious drawback to adequate servo response which will affect the machine's performance and accuracy.

3) Consider the overall height of the orbiting unit versus the Z-axis "window" of your machine. If your machine has a maximum Z height of say, 12-inches, and your orbiter has a thickness of 4-inches, your machine is now limited to accepting parts 8-inches tall, minus vises, tooling, etc. (Also, don't forget to allow for the length of your electrodes.)

4) Be selective. These things are expensive, both in their purchase price and what you have to lose if you don't choose wisely. Ask the salesperson for a list of users in your area and consult them. Don't be shy — few EDMers are. The old saying, "Ask the person who owns one," certainly applies here. The owner is either happy and making money with the machine, or is unhappy and/or losing money, and who better to tell you about its good and bad points. Ask about its accuracy and repeatability, ease of set-up and use, maintenance, the quality of service from the dealer or company, and the length and scope of its warranty.

SUMMARY

In closing, equipping a manual EDM machine with a properly matched, high-quality orbiting system is probably the single greatest performance enhancement possible, second only to a high-quality, basic tooling package. An orbiter used correctly will:

1) reduce machining times;
2) reduce electrode wear;

3) reduce raw material consumption;
4) improve surface finishes;
5) improve accuracy;
6) reduce scrap; and
7) reduce operator involvement.

If you have manual EDM machines and haven't already done so, you should consider investing in an orbiter for your shop. Time and money are among our rarest of commodities, and an orbiter will help you save both of these.

Chapter 6

CNC OR MANUAL

INTRODUCTION

For many years, EDM was considered a "black art" or "much too difficult" by most of the manufacturing community, and therefore was carefully avoided, especially by the larger companies. This new process was taken up largely by the smaller tool and die and specialty shops. The nonusers who *did* require EDM services had to settle for whatever delivery and price they could get. These were not always in their favor, further alienating this already skeptical group.

As generators and controls became more refined and applications became more obvious, EDM became a mainstay especially in the moldmaking and tool-and-die industries. The larger manufacturing firms were busy embracing the new, conventional, high-speed NC and, later, CNC "chip-cutting" machine tools. When CNC EDM was first introduced it was slow to be accepted, even by manual EDM users.

Comments such as, "Where are the crank-handles?" and, "How can a computer come between my hands and my machine?" or "We don't do production. We only do single cavities so we would never need one of those," were often heard during tool shows and sales presentations. This resistance was only natural.

Later, as speeds, finishes and practicality were being proven by the few who were bold enough (and moneyed enough) to make the transition, general acceptance became more commonplace. Now, except for certain applications or operations, competing against modern, properly tooled CNC EDM equipment by manual EDM machines is very difficult. Let's examine why.

FEATURES

The modern power supplies and controls for today's EDM are incredibly advanced, even when compared to the latest CNC machining and turning centers. Besides all of the usual computer-aided capabilities found in conventional machining operations, CNC EDM can provide these and other features inherent only to EDM.

With the advent of CNC-controlled EDM equipment, a whole new realm of features became commonplace. EDM controls offer many of the same features that were already available on conventional equipment: mirror imaging, scaling, pattern rotation, etc., plus many more features that conventional machine users haven't even dreamed of. But, why would we need all these bells and whistles?

Whether your application is in aircraft/aerospace, moldmaking, production, or prototype, CNC EDM can significantly increase your company's efficiency, quality, production, and — needless to say — profit.

Whether you are CEO of a multinational corporation, or the owner of a "mom and pop" job shop, it is common knowledge that the single greatest cost of any operation is labor. To reduce these costs, the reduction of direct labor is required. The concept of CNC EDM is designed with this objective in mind. So, if you could realize increased production, fewer rejections, and cut direct labor costs by *as much as two-thirds*, would you consider it? I am confident you would.

Unattended operation is CNC EDM's forté; and to support this premise are the sophisticated controls and power supplies that are presently in use and are constantly being improved upon. The following examples are just a few of the innovations that are available on most modern CNC EDM machines.

CONTROL AND GENERATOR FEATURES

1) **Adaptive Power Supplies**. Modern power supplies have the ability to monitor or "read" the conditions within the spark gap and automatically make the necessary adjustments to resume or maintain stable machining conditions. This feature allows for the automatic extension or "stretching" of the off-time during marginal machining conditions. When the machine encounters difficulty with poor flushing, deep ribs, etc., the control can extend the off-time to allow more time for reionization of the dielectric. An added advantage to this feature lies in the fact that this occurs *only when necessary*, and *automatically*.

   Many manual EDM's are not this sophisticated. This situation would require the ongoing presence of an operator to constantly adjust and compensate for the marginal conditions as they change or to initially set the control to a safer, but less efficient, duty cycle. This method results in slower machining times and increased costs.

2) **Mirror Finishes**. The advanced circuitry provided in modern power supplies can significantly reduce the recast layer, an important engineering concern of many EDMed applications, especially in the aircraft and aerospace field. The submicron finishes obtained by today's CNC controls can reduce or eliminate entirely the secondary operations often required to remove the recast material. Metallographic studies in ferrous alloys have proven this to

be 0.0004" or less, requiring little or no further processing. Again, any time you can reduce or eliminate a secondary operation, you save labor costs, time, and unnecessary handling. All of these factors will eventually show up in the profit margin.

Figure 6.1   Near-optical quality finishes can be obtained with modern equipment. This is a contact lens mold with electrode.

3) **Automatic Pick-Up Routines**. This feature allows the operator to automatically edge-find or center-find a part. This can be done with a touch-probe used specifically for this, or the electrode itself. Canned location routines can provide internal-center, external-center and corner-datum locations within tenths, totally unattended. More time and labor savings.

4) **Automatic Jump Control**. This same adaptive process applies to a programmed "jump" or retraction of the electrode in the opposite direction of the axis being machined. This is normally done to clear the EDMed area of "chips" and allow clean oil to enter the area. Obviously, when the machine retracts from the spark gap, there is no work being done so we would like to keep this to a minimum.

   Adaptive controls can be programmed to retract the machining axis for a specified amount of time, gradually increasing as the depth increases, but some can also be set to jump *only when necessary*. Manual machines would, once again, require the presence of an operator to either input this jump as required or to set the jump cycle throughout the entire operation, which may not be necessary.

   A CNC EDM is a far more efficient proposition — in many cases eliminating the required presence of an operator. Not that the operator would lose his job but his time could be better spent making and presetting electrodes or workpieces for the next job, or making money operating other equipment that *does* require an operator.

5) **Multiple Work Coordinates**. Another feature that the CNC control can offer is multiple work-coordinate systems. This means that depending upon the machine, you can have up to 100 entirely separate "worktables" or "hand wheels," and one hundred matching digital readout systems. This feature will allow you to have several different jobs or setups in the worktank at once, without trying to keep track of a lot of confusing numbers. This is especially valuable in job-shop situations where work is often interrupted for a hot job or quick repair.

   This feature allows an operator to "leave" the job in work-coordinate system #1, setup and do the "quickie" in work-coordinate #2, then go back to job #1 without "getting lost" or having to relocate the initial setup. How many times has an existing setup been broken or a single-display digital readout (DRO) been cleared to allow for that hot job or emergency?

6) **Mirror Imaging**. Another useful feature is the computerized "mirror imaging" of any or all axes. This useful feature is used to reduce the number of keystrokes required for repetitive positioning programs and, by the same proportions, reduce the possibilities of programming errors.

   The programmer or operator simply programs one side of the centerline and the control will "mirror" the same program to the opposite side of one or both axes simultaneously. Programming time is reduced, and fewer keystrokes by the programmer will further reduce the possibilities of error.

7) **Pattern Rotation**. This is the ability of the control to electronically rotate the program around a given axis. For example, a part with a bolt circle of eight holes located 3" radially from center can be executed by programming the first move and burn at X 3.0" and then "rotating" the program 45°. By repeating or looping this program seven times, the computer will continue to read the X 3" command, but will increment a 45° angle each time and make an X/Y move to the proper location. There will be no need for the tedious trigonometry calculations usually required and, consequently, far less chance for errors.

   Used in conjunction with a C-axis, the spindle can be simultaneously rotated the same angular amount for correct radial orientation of the electrode.

8) **"Thinking" Variables**. This is a programming ability of the control to manipulate numbers within programs to affect desired changes without the presence of an operator. For example, after an electrode has been used, a subprogram or "macro" can direct the electrode to a preset tooling ball. This program is usually a simple edge-finding routine. The numbers are stored internally as variables and compared to the original numbers of an unused electrode, stored previously. If there are any deviations due to electrode wear, the differences can be added to or subtracted from the original program *automatically*. In this way, electrodes that wear progressively shorter by 0.001"

from each cavity will automatically be programmed to advance 0.001" deeper than the previous cavity, resulting in consistent depths and continued unattended operation.

The use of variables allows multiple step math problems to be executed to determine positioning points instead of relying on hand-held calculator input that is prone to human error. Another example where variables are useful is after changing the electrode to a hardened and ground spherical probe and using another linked program, being able to accurately inspect the finished part or cavity just like a coordinate measuring machine. All the measurements can be stored and compared to the "master program" and can be supplied as a computer printout with the finished part. By the way, most modern vertical CNC EDM machines are much more accurate than most coordinate measuring machines.

OTHER CONTROL FEATURES

So far, we have examined only a few features that are built-in or options on today's modern EDM controls. Touching on a few more are:

1) **Conversational Programming**. This allows even entry level operators to program the machine without knowing or learning G- and M-codes.
2) **Canned Orbiting Cycles**. Many useful 2-D and 3-D orbiting patterns and shapes are resident in the machine and can be easily called up and used or edited.
3) **Axis Exchange**. The ability to reorient the workpiece (let's say from flat on the table to vertically against an angle plate) without rewriting the program.
4) **Scaling**. This allows program dimensions to be scaled up or down. This is very useful to moldmakers. For example, it is possible to program a 1" cavity then introduce 16% material shrinkage.
5) **Estimating Screens**. Allows time studies and cost projections to be done right at the machine. This information is usually available after the part is described to the control.
6) **Machining Status**. Many machines provide the ability to track and chart machining performance using visual tables and graphs. This info can be downloaded to a PC.
7) **Preventive Maintenance**. Most CNC EDM's provide diagnostic screens that display information on filter life, cumulative machining time, lubrication intervals, etc.

The previous examples are on the control side and are software related. There are many hardware features of the machine-tool itself that lend itself to productivity and efficiency, ranging from semiattended operation to complete machine autonomy.

Let's examine a few.

MACHINE FEATURES

1) **Programmable Flushing**. The machine can be programmed where and when to use pressure or vacuum flushing, but more important, it can turn on and off multiple flushing lines as they are needed. This is very useful when machining multiple cavities or details that are being flushed through the part instead of through the electrode.

2) **Spindle Rotation**. A very important capability that can be supplied with most modern CNC EDM's. This can be an aftermarket add-on or supplied with an integral rotating spindle. Rotating spindle speeds range from 0 to 2,000 rpm and can be manually set or preprogrammed through the control.

   Spindle rotation is useful for machining uniform, round openings and through-holes. This aids in flushing and in most cases will provide more uniform electrode wear.

   Another useful application for spindle rotation is discharge dressing of round metallic electrodes or vertical turning of graphite electrodes. This is covered later in this chapter and in more detail in the "Electrode Rotation" section in Chapter 8.

3) **C-Axis**. There are many features a CNC EDM machine offers that manual machines can't address at all. By using the programmable C-axis, the EDMing of helical gears, worms, threads, and turbine vanes is easily accomplished. Indexing of the electrodes or parts unattended and with CNC accuracy is another obvious advantage. This feature is presently part of an integral EDM process used by many plants engaged in jet engine, turbine, and turbo-machinery manufacturing. This is also a very popular feature within the moldmaking industry.

4) **Automatic Tool Changers**. One of the single largest steps toward full automation is equipping a CNC EDM with an automatic tool changer. These devices range from smaller 4-station shuttle-type changers to 48-station carousels capable of changing electrodes *and* workpieces. This is the single-great-

Figure 6.2   On machines without an automatic tool changer, electrodes can be mounted on a turret-type holder and indexed as needed using the C-axis. (Courtesy of Sodick, Inc.)

est contribution to providing a shop with continuous, reliable, unattended overnight operations.

5) **Automatic Workpiece Changers**. For the progressive operation that desires the utmost in autonomy and efficiency, workpiece changers are available in many configurations including robot-like devices to transfer the workpieces or shuttle type pallets sliding in and out on a rail system. They are often used in conjunction with an automatic tool changer.

CAPABILITIES

We have examined a few of the control and machine features now available, many of them standard on modern CNC EDM's. These features, when combined, offer an almost unlimited degree of capability. For the most part, anyone seriously intent upon fully exploiting the capabilities of CNC EDM would be limited only by their imagination and ingenuity. Please consider the following.

1) **Multiple Spindles**. With CNC EDM, we are not limited to a "one spindle, one cutter" mentality like conventional equipment. With EDM we can mount the part in the spindle and place multiple electrodes on the worktable. By reversing the polarity we can effectively "trick" the machine into thinking it is "multi-spindled."

2) **Automatic Part/Electrode Qualification**. The CNC control can also allow us to probe each electrode or workpiece and record and save its location for precise machining at each station. It can also inspect its own finished parts by following an inspection program with a probe and then recording the results in a manner of a DCC coordinate measuring machine.

3) **Low Machinability Alloys**. In aircraft and aerospace applications — with low machinability alloys and tolerances that keep tool and cutter suppliers very happy — EDM is often the logical alternative to conventional machining. Inconel is hard on cutters and equipment, as is titanium, Waspalloy, or even worse, MP-35. The workpiece material's hardness, detail, shape and finish can severely limit conventional machining. Other factors that often eliminate conventional machining practice such as material work-hardening, cutter-whip, undercutting, tool wear, dimensional tolerances, and burrs can also enter the picture.

4) **Simplify Parts Setup**. For example, a prehardened hydraulic sleeve, used in the aircraft and aerospace industries, can have tolerances of ± 0.002" on corner radiuses and ± 0.0001" co-planer tolerances on all metering edges. These can be fixtured on a precision arbor and mounted vertically in the spindle of the machine. By positioning multiple electrodes horizontally on the worktable and using the C-axis for indexing, dozens of labor intensive setups can be eliminated. Besides reducing the time required and the consequent labor costs, this greatly reduces the chance for costly errors that are possible with repetitive, time-consuming manual setups.

5) **Parts Changer**. Another cost-saving capability often overlooked is the ability to use the automatic tool changer as a parts changer or parts carousel. In normal EDM operations, the part is located on the worktable and the electrode is in the spindle. Instead of changing to another electrode when the preceding electrode is worn or its particular operation is completed, the part is simply presented to the next electrode that is already prepositioned in the worktank. Instead of making multiple electrode changes using the tool changer, only a single "tool change" is required. The "tool change" is when a finished EDMed part is returned to the parts carousel and a blank part is loaded and presented to the electrodes. Instead of executing say, eight mechanical electrode changes per part, this method reduces it to only one, presenting one part to eight electrodes.

In a traditional set-up (workpiece on the worktable), when the part is finished, the pump must be turned off and the tank drained. The part is then changed, the pump must be turned on, and the operator must wait for the tank to fill to a safe level. If this emptying and filling cycle takes longer than a few minutes, it can cost a lot of money over the life of a long-run production job. Even if it takes only one minute to empty the tank and one minute to refill, every 100 parts will add *over 3 hours* of "empty and refill time" to the machining times.

By reversing polarity and mounting the parts on holders in the tool changer and placing the electrodes on the table, the tank never has to be drained until the "gang" of palletized electrodes needs replacing. All that is required of an operator is to remove the finished parts from the tool magazine and replace them with premounted blanks. The operator is not out of a job. He can be presetting blank parts and making or redressing electrodes during the machining cycle. Fast, efficient, and profitable.

This method also improves flushing. Since the cavity is being machined up-side-down, the chips will want to fall out of the cavity instead of down into it. This should reduce machining times, improve finishes and minimize the potential for arcs and pitting.

6) **Moldmaking**. A CNC EDM machine equipped with an automatic tool changer, programmable flushing, and an integral rotating/indexing spindle provides an indispensable machine tool for the mold making industry. This combination is ideal for the multi-cavity applications common in this discipline. In addition, the mirror finishing circuitry available in most machines today can provide a "press-ready" cavity finish that can eliminate or substantially reduce costly polishing operations that are expensive, labor intensive, and could actually damage sharp edges and fine detail.

Traditionally, most blind cavities are EDMed in simple "Z-minus" type operations that require multiple roughing and finishing electrodes. This process yields excessively long machining times, less than desired finishes, and unacceptable electrode wear — not to mention the time, labor and equipment required to fabricate numerous electrodes of different sizes. Since there is nothing "traditional" about CNC EDM, you need not be "traditional" in your thinking. Remember, your only limitation is your imagination.

An effective "CNC Approach" would be to rough and finish the cavity with a single, undersized electrode. This is good CNC EDM practice because the entire surface of the electrode is used during finishing operations instead of just the leading edges. Because more surface area is presented to the workpiece, the operator is able to use higher amperages for faster machining while protecting the electrode against excessive wear. This method is much faster, has a better "safety margin" because flushing is not as critical as a non-moving electrode, and it provides better finishes with minimal wear. In almost every case, the savings of time and electrode materials are far more efficient than the "old days" of roughing and finishing with multiple electrodes.

7) **Discharge Dressing**. A further support for unattended or overnight EDM operations is a process called "discharge dressing." This process can be used to redress worn electrodes preparatory to finishing operations *without removing them from the EDM machine*. This is a valuable capability considering that the electrode is redressed in the same spindle it will be used in actual machining. Let's examine this in depth.

With manual EDM, if the electrode is worn, the machine must be stopped. Now you have a high-dollar piece of equipment — the machine — "idling at the curb." Next, the electrode must be removed from the machine, and now you are tying up another piece of high-dollar "equipment" — the operator. Now, depending upon its availability, he must tie up yet another piece of high-dollar equipment — the lathe, mill or grinder required to redress the electrode. All of this overhead is involved in supporting an EDM machine that, for the moment, isn't earning a cent.

Now, let's examine the "CNC Approach." With CNC, when the electrode needs redressing, the control directs the machine to withdraw the electrode from the cavity and "send" it to a pre-positioned "dressing block" located elsewhere in the worktank. By "tricking" the machine to wear out the *electrode* instead of the workpiece, we can dress the electrode in a matter of minutes, compensate for the amount we removed, return to the main program, reset the zero in the main program, and resume the machining operations — *all unattended!* There will be no tying up a lathe, mill or grinder, no tedious setups or reindicating the electrode at the support machinery or in the EDM machine!

Granted, the EDM machine was not actually earning any money while it was dressing it's own electrode, but neither was it costing any money by tying up your operator or additional support equipment. This can free them to do more important things — like making money doing something else! Their time can be better spent doing something far more productive than babysitting a manual EDM machine. If the enormity of this capability has not convinced you, try to imagine your savings in time and money if your milling machines could resharpen their own end mills and you always had a "finisher" ready! We will take a closer look at these savings later in this chapter in the section on "Cost Justification."

8) **User Friendly**. Finally, to overcome the sometimes natural resistance to CNC EDM that entry level users may have, programming software is often provided as a standard feature. In many cases, the operator does not need to know much about EDM technology *or* CNC programming. In a conversational mode, the computer/control is told the size and type of electrode material, and the size and finish of the workpiece. Almost instantly the computer can select power settings, orbit steps, write the program, and in some cases, estimate how long the job should take! Today's shop-owner does not always need "high-tech" operators to operate this high-tech equipment.

ADVANTAGES

A few important advantages of CNC EDM are:

1) **Applicability**. CNC EDM offers the ability to successfully machine the formerly impossible manufacturing challenges, whether by innovative approaches toward the end results or the inherent capabilities of the EDM process itself.

2) **Flexibility**. No other process allows the manufacturer the flexibility that CNC EDM does. The ability to "switch ends" and hold the workpiece in the spindle is just one example. "Sharpening" its own "cutters" (discharge dressing) is another.

3) **Reliability**. Coupled with high-speed microprocessors and adaptive power supplies, CNC EDM offers a very high degree of machining reliability. In addition, many machines come standard with integral, self-checking, diagnostic systems.

4) **Accuracy**. Positioning and repeatability are in the range of 10 *millionths*. Very few, if any, other machine tools can boast of this degree of accuracy. This will also ensure good machining results because of the sensitivity of the servo system to recognize and maintain the correct spark gap.

5) **Profitability**. All of the above features, coupled with the time- and labor-saving advantages provided by unattended machining, virtually assures any firm with sound planning and good "machine shop practice" increased production, improved quality control, and significant cost reductions in EDM operations.

As mentioned much earlier, direct labor is the single greatest part of a company's overhead. Without question, a CNC EDM system will reduce the amount of direct labor that is required for practically any application conceived of, not to mention making the formerly "impossible" jobs a matter of routine.

All of this sounds good so far, but how the heck do I pay for something like this?

COST JUSTIFICATION

Probably the single greatest hurdle for an interested party in purchasing a CNC EDM is the fact that *none* of them is inexpensive — with or without all the bells and whistles.

So, how can the small job shop justify the capital expenditure for a single machine? On the surface, it can seem difficult for a small shop to be able to justify the major expense of a CNC EDM, but when you examine all the possibilities it introduces, it can make this justification almost easy. There are many methods to help justify a capital expense but the bottom line in any case is your *return on investment*, or ROI. This can get pretty complicated for a small job shop unless a large enough contract is obtained, and a fixed income can be relied upon for calculations. Since most small shops do not have the luxury of one or more long-term contracts, we will look at it from a jobber's point of view.

Suppose it is Monday in a single-shift shop and we have a 16-cavity injection mold to build. Nothing fancy, perhaps cavities of 1 inch square and 0.250" deep, with a bottom corner radius of 0.010" maximum. For this comparison, the setup is made and electrodes are ready.

The chart in Figure 6.3 is based on a shop rate of $50.00 an hour.

CNC EDM			MANUAL EDM	
Time	Cost	OPERATION	Time	Cost
45 min.	$37.50	ROUGHING	45 min.	$37.50
10 min.	8.33	FINISHING	20 min.	16.67
5 min.	4.17	DRESSING	5 min.	4.17
N/A	N/A	FINISHING	10 min.	8.33
60 min.	50.00	1 CAVITY	80 min.	66.67
X 16	X 16		X 16	X 16
16 hours	$800.00	16 CAVITIES	21.33 hours	$1,066.72

Figure 6.3   Comparative costs of CNC and manual EDM.

In addition to the direct, hourly savings of approximately $266.70 in machine-time, the CNC method was completed in only two *shifts* instead of two and a half *days*. This means that job #2 can go into the CNC EDM on Tuesday morning while job #2 can't be started by the manual machine until sometime Wednesday afternoon. In this application, unattended CNC EDM operations (including discharge dressing the worn electrodes) can free the machine for the remaining four work days with a total earning potential at shop rate of $1,600.00 per shift versus only 2.4 days remaining for the manual machine and a reduced earning potential of $933.50. So on Monday we saved $266 in machine-time and still had the third-shift open with a potential of $400, a $667 potential income increase on the first day! Operating in this manner day after day, it is clear that the CNC EDM will continue to increase the difference and the manual shop can never catch up.

One must also consider that electrode costs for the CNC will be less because fewer electrodes will be needed and they will all be the same size. Nor will a grinder have to be set up and remain "on-call" for timely redressing. If you add the cost of a surface grinder remaining on "stand-by" for two days and the expense of an operator to redress the electrode each time, this can increase costs an additional $900.00 — in effect more than *doubling* the cost of the CNC method.

At the time of this writing, the average selling price of a new, vertical CNC EDM is approximately $140,000. To purchase this machine, the monthly payments over 5 years at 8.75% interest will be $2,890. If this machine runs daily for only one 8 hour shift, its potential gross is $8,600 a month. (This amount can be much higher because, in most cases, this machine will run overnight many times a month). This number does not include the hidden savings realized by the operator being able to work on other jobs while the EDM is busy. The operator could be busy preparing electrodes for the next job, or could be working on the mold base while the EDM machine is finishing the cavities. Nor does it include the reduction or elimination of secondary polishing operations that involve time, labor, and necessary polishing equipment and supplies.

Ok. What's the bottom line? The bottom line is one word: *Throughput*.

Throughput means: The more work that crosses a machine's worktable every hour, day, week or month, the greater the increase in that operation's productivity and, not coincidentally, their profitability!

THIS SOUNDS REALLY GOOD BUT. . .

I agree. All of this sounds wonderful *if* a shop has a steady amount of work to provide this kind of income. Even if you don't feel you have enough work to support such a venture, a CNC EDM can actually help increase the volume of work your shop can provide by allowing discounted shop rates for "lights-out" EDM services. This will attract a higher percentage of quoted jobs.

For example: There are ten EDM shops on a given street, five on each side. They are all dependent, for the most part, on the large manufacturing firm at the end of the street. All the job shops have similar equipment and have the same shop rate of $50.00 an hour. Considering all the bidding is comparable and the buyer "takes turns," how can any one shop get and maintain an edge over the other nine shops?

The answer is: *Unattended operations using CNC EDM*. Instead of sending in the next quote with your usual $50.00 an hour rate and taking your chances with everyone else, reduce your rate to $40.00 an hour and run the job overnight. The buyer will be astounded at the fact your bid is 20% less than the rest of your competitors; and since your previous quality records have been good, chances are high you will be awarded the job.

Granted, you *are* taking a 20% cut in your shop rate but only during the day shift. You make up this shortfall and more on unattended second and third shifts. That's $40 an hour coming in that *you weren't earning before and* you earned them *while you were sleeping!!* When was the last time you made money while you were sleeping?

Results? Your product is CNC accurate, your quality is high, your deliveries are timely, and you are getting more work volume across your machine's worktable than ever before. You are happy, the buyer is happy, your customer is happy. *Everyone* is happy except your competitors.

CONCLUSION

All of the features and capabilities that have been described here will translate into significant savings of time and far more efficient use of manpower and machinery. If these are *not* concerns of yours, then I must apologize for taking your valuable time, but somehow I have the feeling that if you have taken the time to read this far, you *are* interested in reducing costs, increasing productivity, improving quality and ultimately increasing the profitability of your operation. Once you are using CNC EDM, you will discover the numerous other time- and labor-saving features and capabilities that it offers (that we don't have space to discuss here).

The EDM process has been around for almost 50 years and is no longer a black art or a mystery area where manufacturers lose jobs and/or money. Now, augmented by the powerful capabilities that modern CNC can provide, EDM has rightfully taken its place alongside lathes, mills, and grinders as full-fledged profit centers, and has become an integral part of the modern manufacturing process.

Just remember. . . When it comes to CNC EDM, *you're only limited by your imagination!*

SUMMARY

Now, I realize that the discussion in this chapter may bother some of the people who still prefer manual machines — and especially the people who are trying to sell them — but this has been objectively written to show the capabilities and advantages that CNC EDM has over manual equipment.

Granted, some jobs or operations don't warrant a full-blown CNC-controlled EDM. Simple hole drilling, occasional detail or cavity burning, or jobs requiring extended times "down-in-the-burn" don't necessarily warrant the extra cost of a high-end machine.

But most of the jobs coming down the pike today are of the high-end nature, such as multicavity, machine-polished injection molds, hydraulic sleeves, spools and manifolds, or close-tolerance production. Most of these applications would require unattended machining, adaptive controls and automatic tool changing to remain competitive.

In addition to all of the so-called "standard features" that a CNC EDM can provide, there are companies and individuals that may have a truly special application or requirement that calls for something out of the ordinary. This type of equipment affords the creative designer or progressive-thinking company the ability to *go outside the realm of conventional thinking,* and allow the flexibility to create or modify a machine to be capable of doing far beyond what it was initially conceived to do. Progressive, imaginative mentalities would think like this:

"We can add a servo-motor here, extend this side of the tank, hold the part upside down...."

It is with *this* kind of thinking and imagination that allowed lightbulbs to be invented, and airplanes to fly.

Chapter 7

DIELECTRIC OIL

INTRODUCTION

In previous chapters we examined the three basic types of flushing methods used with vertical EDM. Then, we covered several advanced flushing and trouble-shooting techniques. At this point we will examine the flushing medium itself — the dielectric fluid or EDM oil. What is it made of? How do we know if it is working well? What about operator health and safety?

Even if we could master all flushing techniques (if possible) or had created one of the most imaginative and efficient flushing setups in the world, it wouldn't mean much at all if, because of an inferior dielectric fluid, we burned the shop down or all the operators were absent because of dermatitis (or worse). We need to know more.

EDM DIELECTRIC FLUID

Oils have been used as dielectric fluid as long as the process has existed, but only in the past decade have any appreciable research or scientific approaches been made as to their composition and compatibility with people and the environment. Recent health and safety concerns have accelerated this overdue interest.

There are many different types of fluids available from EDM suppliers and refiners. Typically, fluids with paraffinic, napthenic, and aromatic bases are used. Even though I am not a chemist, I strongly recommend that you use only dielectrics that have been produced specifically for EDM use. Many dielectrics that have been sold by nonspecialty refineries have actually been byproducts from other processes that "just happened to work" for EDM and were sold as such. You must choose a dielectric fluid that will be compatible not only with your EDM operation but, more important, with your EDM operators. Just because "it works" doesn't necessarily mean that it is safe. This won't make your choice any easier, but it will narrow your field of choice considerably.

Again, be very selective in your choice. Even after you make your selection and the oil is delivered, inspect the barrels thoroughly. Several years ago, while managing a large mold shop, I witnessed the delivery of several drums of EDM oil. Upon closer examination, a skull-and-crossbones emblem was found stenciled on the side and the label below it read: "*Caution! Prolonged breathing of vapors or exposure to skin can result in kidney damage!*"

Ahem….Well, the last time I checked, the only EDM operators that are *not* subject to prolonged breathing of these vapors or exposure to skin are either unemployed or retired! Operators breathe the vapors from dielectric fluid daily and have their hands in it constantly. The truly disappointing part of this situation was that this oil was produced by a major refinery, and supplied by a major supply house. Both companies must have been illiterate or irresponsible (or both!). Hopefully this condition no longer exists although, unfortunately, mineral seal oils are still used and are, without exception, carcinogenic. As for the EDM oil in question? We refused delivery and sent the drums back to the supplier. Having learned our lesson on price-buying, we took the time and learned quite a bit more about EDM oils. Eventually, we wound up spending more than we had initially intended, but we purchased a quality oil and peace of mind.

MATERIAL SAFETY DATA SHEET

Use no EDM fluid that doesn't come with an MSDS (Material Safety Data Sheet) or isn't supplied to you on request. An MSDS is a form outlining results of a product's safety after undergoing testing set to industry wide standards. This includes OSHA for the United States, and WHMIS for Canada.

The details of the MSDS form are explained in Figure 7.1.

Don't allow the technical language or wording of the MSDS form to frighten you, but since this is the same form that is used for many other products and chemicals that may be highly poisonous, caustic, or radioactive, you must understand the necessity for such specific and business-like language.

EDM OPERATOR CONCERNS

To keep it simple, the following list explains the things that the EDM operator should be concerned with.

Flash point. This is the temperature at which the vapors of the fluid will ignite. This explanation is a bit simplistic as conditions for testing are more involved, but for the sake of discussion and for safety's sake, the higher the temperature, the better. Unless you are doing extremely small, low power cavity work, or drilling the tiniest of holes, be especially concerned with anything on a spec sheet or MSDS rated lower than 82°C (180°F).

1) **Material Identification and Use** — This identifies the specific name the product is sold under, the name and address of the manufacturer, and what its intended use is.

2) **Hazardous Ingredients or Materials** — This section identifies any hazardous ingredients (if any) and describes their concentration and lethal doses.

3) **Physical Data for Material** — Defines whether the product is a solid, liquid, or gas; odors; specific gravity; boiling and freezing points; evaporation rate; etc.

4) **Fire and Explosion Hazard** — Indicates if the product is flammable or explosive and if so, under what conditions. Also outlines procedures to follow in the event of a fire, including means of extinction and what kinds of smoke, fumes, or gasses would be given off.

5) **Reactivity Data** — Indicates product's chemical stability or incompatibility with other substances.

6) **Toxicology Properties of Product** — Lists all routes of possible product entry into the body and the effects of acute and chronic exposure to the product, including lethal dose, if any. Also lists irritancy, exposure limits, and whether product is carcinogenic, teratogenic, mutagenic, or affects reproductive functions.

7) **Preventive Measures** — Outlines protective personal gear to be worn (if any) when handling, storing, or disposing of product. Indicates leak and spill procedures and ventilation requirements.

8) **First Aid Measures** — Outlines first aid procedures (if required) in the event of contact with skin or eyes and inhalation or ingestion.

9) **Preparation Data of MSDS** — Shows name, address, and telephone number of the testing agency, laboratory, or preparer. Also shows the date of document preparation.

Figure 7.1   The nine disclosure sections of an official MSDS form.

Dielectric strength. This is the fluid's ability to maintain high resistivity before spark discharge and, in turn, the ability to recover rapidly with a minimal amount of off-time. An oil with a high dielectric strength will offer a finer degree of control throughout the range of frequencies used, especially those when machining with high duty cycles or poor flushing conditions. This will provide better cutting efficiency coupled with a reduced potential for arcing.

Viscosity. The lower the viscosity of the dielectric oil, the better accuracies and finishes can be obtained. Sometimes referred to as "gap penetration." In mirror finishing or close tolerance operations, spark gaps can be 0.0005" or less. With such tight, physical restrictions such as this, it is much easier to flush small spark gaps with a lighter, thinner oil.

Good "finishing" EDM oils are on the thin side. When EDMing applications that require high current levels and high metal removal rates, or those requiring only moderate finishes such as those found in forging dies, heavier oils can be used. Viscosity can be heavier in these applications because of the naturally larger spark gaps, and it will also help prevent excessive fluid loss through vaporization.

Specific gravity. This is often confused with viscosity. It is the actual "weight" of a substance measured by a hydrometer. The "lighter" the oil (or the lower its specific gravity), the faster the heavy particles (chips) that are suspended within it will "settle out." This reduces gap contamination and the possibilities of secondary discharge and/or dc arcing.

Color. All dielectric oils will eventually darken with use, but it seems only logical to start with a liquid that is as clear as possible to allow viewing of the submerged part. Clear or "water-white" should be your choice, because any fluid that is not clear when brand new, almost certainly contains undesirable or dangerous contaminates. Ask for and review the oil's MSDS.

Odor. Besides the obvious aesthetic reasons for choosing a fluid with no discernable odor, oils that "smell" are usually a strong indication of the presence of sulfur, which is undesirable in EDM oil. A well-known authority on EDM fluids has often said, "Just because you are an EDM operator doesn't mean you have to smell like one." I'm sure that most of us (and especially our spouses) would agree.

PREVENTIVE MAINTENANCE

Depending on its use and maintenance, dielectric oils can last for several years. Regularly scheduled filter changes and prevention of water contamination will considerably extend their useful life.

Water contamination cannot be avoided entirely as condensation will occur on the electrode surfaces as it heats up. Graphite electrodes will contribute more to this condition than metallic electrodes, as their porous structure will actually attract and absorb moisture from the air. Storing graphite electrodes in a clean, dry area will keep moisture absorption to a minimum. Some shops using very large graphite electrodes will place them in a low-temp drying oven the night before their intended use (but this is an exception, not the rule). A less obtrusive method to keep humidity and moisture absorption to a minimum would be to allow a 60 watt lightbulb to remain lit within the storage cabinet.

The color of the oil is not necessarily an indicator for oil replacement. All oils, no matter how clear they are when new, will darken in shades from amber to brown with use and age, because these products will break down when exposed to high heat. Obviously, sustained high amperage machining will break the oil down more rapidly. Tars, resins, and hydrocarbons are generated when this occurs, and this is what "stains" the oil. No amount of filtration will remove this discoloration, so don't mistake "colored oil" for "dirty oil."

Dirty oil is best judged by several factors.

1) Pressure gauge readings as described in the machine's maintenance manual.

2) Increased occurrence of dc arcing or pitting with settings that were previously successful. (Assuming, of course, that no other changes have been made, such as the grade of the electrode material, flushing pressures, etc.)

3) Longer cutting cycles and/or degradation of finishes.

4) A visual inspection of the oil. To do this, fill the work tank but do not machine. Collect a sample of oil in a clear container. Visually check the sample immediately for any cloudiness, and again after a few hours to check for sediments on the bottom or color striations in the oil itself. Dirty oil is usually tinted gray or black, and this coloration will dissipate if the oil is allowed to circulate through the filtration system. Replacing the filters is the least expensive and the most likely remedy to the above symptoms.

If a filter change reduces pressure gauge readings, but does not correct the deteriorating cutting conditions, then replacement of the oil may also be warranted. Generally speaking, unless the machine uses only very low power settings and/or high frequencies, it is safe to consider replacement of the oil after 18 months to 3 years of daily use.

Replacing your EDM oil can be an expensive proposition, especially in larger machines and central filtration systems. To be absolutely sure when this is necessary, there are several firms that can test a sample of your oil for acidity, hydrocarbons, water, oxidation, and other contaminants, and they can advise you in a much more scientific manner than I have been able to do in this chapter.

In the event an oil change is required, drain the reservoir completely, and remove all sludge and sediment. Wipe the greasy film from the sides of the sump and inside of filter canisters before refilling with oil and replacing filter elements. (*Note:* Check with local authorities concerning the correct methods of waste oil and filter disposal.)

If the dielectric system is exceptionally dirty or has not been used for several months, partially fill the sump with just enough fresh oil to prevent the pump from losing prime, and allow to circulate for approximately one hour to dissolve and wash away any films or coatings remaining inside pipes and hoses. Most EDM oils contain some active solvents (especially when fresh), and will help clean the inaccessible areas of the dielectric system. Drain this oil and dispose of properly.

Refill the reservoir with a high-quality EDM oil made especially for this purpose. Some shops use deodorized kerosene with success, but keep in mind this is not a product that has been engineered specifically for EDM use. Those who use it like its low viscosity and finishing capabilities, but it has many undesirable qualities such as low flash point, high evaporation rate, and it will cause excessive defatting of the skin, resulting in irritation, redness, and possibly dermatitis.

Caution: Due to its low flash point, deodorized kerosene may not meet city or state fire codes and other regulations in your area. If you are already using it, check with your local authorities concerning its safety and legality. If you are considering using it, don't.

Again, I strongly recommend using dielectrics that have been engineered specifically for EDM use. If you have difficulty in your oil selection process, call the machine manufacturer or an EDM supply house for their recommendations.

SUMMARY

Beyond the obvious safety concerns, these guidelines contribute to productivity and profitability. Purchasing an inexpensive but inferior EDM fluid can cost you much more than the difference in price for a premium oil. Oxidation, viscosity, dielectric strength, etc., are all factors that will affect future productivity of your operation and, in every case, decreased cutting speeds, pitting, increased dc arcing, and filtration problems will accumulate and ultimately cost you far more than any "savings" you thought you realized.

Successful EDMing begins in the planning stages. Flushing provisions are best made in the design stages of the mold or part itself and in the electrode design. We don't always have this luxury, but try not to make flushing an afterthought — your project will suffer if you do.

As for EDM oils, we have all heard the old adage, "flushing, flushing, and flushing," but you must first know and understand the characteristics of the fluid you are using. All of your careful planning of electrode design and flushing provisions can mean very little if you are not using a premium grade, safe EDM fluid.

Quality dielectric fluid and flushing go hand in hand. Know them both or you could be flushing health and profits right down the drain!

Chapter 8

DISCHARGE DRESSING ELECTRODES

INTRODUCTION

Discharge dressing is not a new process, but for whatever reason it is not in common use, although it is an incredibly powerful enhancement in support of unattended operations. Discharge dressing can be performed in a manual EDM machine, but this process is best suited to a CNC EDM. The reasons for this will become clear as this chapter develops.

THEORY

Normal EDM operations strive to *protect* the electrode, holding wear to a minimum while eroding the workpiece. This is achieved by using strategic combinations of electrode polarity, current, and frequencies.

Deviations from this "ideal" occur not just during discharge dressing operations but when finish, surface integrity, machining speed, workpiece material and other conditions warrant, but for the most part, we would choose to protect the electrode whenever we can.

Discharge dressing is a process where we depart from convention and intentionally "wear out" the electrode. The concept is simple: adjust the EDM control settings to provide the highest wear conditions possible: negative polarity, excessive amperage, high frequencies, and capacitance.

Any EDM operator with even a minimal amount of hands-on experience knows that these conditions contribute to high rates of electrode wear, so these conditions are usually avoided whenever possible — with a few exceptions.

ELECTRODE WEAR —
ALMOST NEVER DESIRABLE

Some workpiece materials *require* the use of "wear-settings." Wear-settings are used when EDMing carbides, titanium, metals in the copper family, and refractory materials. With these materials and a few others, no other method is successful, so we must accept high electrode wear.

Another application that would require the use of "wear settings" is when machining *speed* is more important than electrode *wear*. This can be when the electrode shape is simple and easily replaced or redressed, as in the case of very simple geometric shapes or when tubing a hole. Other applications could be when roughing larger cavities with a low-grade graphite, or when electrodes are to be quickly and easily redressed by abrading. In these examples, we desire only speed and consider the electrode expendable.

Discharge dressing comes to play whenever we wish to make or redress an electrode in the EDM machine. Copper electrodes lend themselves very well to discharge dressing, while graphite yields almost poor results. This difference in dressing performance is based solely upon their physical characteristics. We will examine these now.

WHICH ELECTRODE MATERIAL?

Copper will readily discharge dress with very few problems. Its relatively low melting temperature of 1082°C (1980°F) and high conductivity are the greatest reasons for its success.

In comparison, graphite is very low in conductivity and does not melt at all, but *sublimates*; that is it passes directly from a solid to a gas, bypassing the liquid state entirely. This occurs at a much higher temperature of approximately 3350°C (6062°F). Graphite's extremely good resistance to thermal damage is the reason it resists wear during machining and, for the same reason, has poor discharge dressing capabilities.

Discharge dressing graphite is about as easy as taking the high-speed steel end mill from the spindle of your milling machine, clamping it in the vise, and trying to machine *it*. Like an end mill, EDM graphite is designed to *machine*, not *be machined.*

Conclusion: We will *almost never* discharge dress graphite electrodes; only copper.

There are other tricks we can do to dress graphite electrodes in the machine that we can't do with metallic ones, and these will be covered later.

PREPARATION AND SETUP

Before starting, we must prepare what is called the *dressing block.* As you become accustomed to doing this, you will accumulate many different sizes and shapes of dressing blocks, depending upon your applications; but for discussion's sake, we will start with a versatile rectangular shape. For this example, let's assume we are preparing to redress the face of a 1" square copper electrode that has 0.010" corner wear.

Purchase a 1" × 2" × 4" bar of 70/30 copper tungsten from your local supply house. Grind all surfaces flat, square, and parallel. Hold this piece in a precision toolmaker's vise with its top surface at least 0.500" above the vise jaws. Locate it in the work tank off to the side of the job you are to EDM. All planes should be true to the travels of the machine within a few tenths (see Figure 8.1).

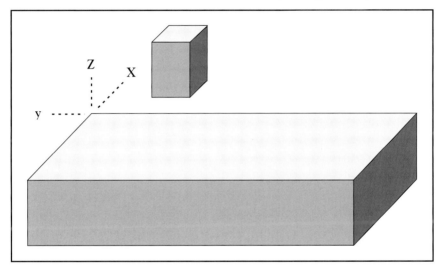

Figure 8.1   Basic setup.

Assuming your workpiece has been setup and is designated work-coordinate #1 (G54, for example), change to work-coordinate #2 (G55) and manually position the electrode over the dressing block. With the electrode, depth-find and locate the top of the dressing-block and set the Z-axis to zero. Manually jog away in the Y-axis until the front face of the electrode is about 0.050" away from the rear edge of the dressing block. Set a zero in the Y-axis. In the X-axis, the entire width of the electrode should be away from the end of the dressing block by at least 0.030". Zero the X-axis. (Neither the X or Y axes locations are critical and can be zeroed "by eye.") This will be the electrode's start-point during dressing operations.

PROCEDURE

For our sample job, we have a 16-cavity mold, with cavities to be 1" square and 0.500" deep, and with a maximum corner radius callout of 0.005" on the cavity bottom. After roughing and finishing the first cavity, we can measure and determine the electrode has a 0.010" radius on the bottom corners. At this point, we want the EDM control to read a subprogram and enter the dressing cycle.

Discharge dressing is accomplished by passing a copper electrode across a dressing block made of copper tungsten. By intentionally wearing-out the electrode and protecting the dressing block, we can shape or redress most simple shapes and quite a few of the more complex ones too.

In this case, the first line of the dressing program should tell the control to switch from work-coordinate #1, and in work-coordinate #2, proceed to the preset locations made during setup, positioning the electrode behind the dressing block with the Z-axis 0.010" lower than the top surface.

Next, the control loads the cutting conditions used for dressing (negative electrode, high-current, high-frequency, and capacitors, if available). The next line should direct the machine to "cut" onto the rear edge of the dressing-block and continue to pass all the way across it (see Figure 8.2).

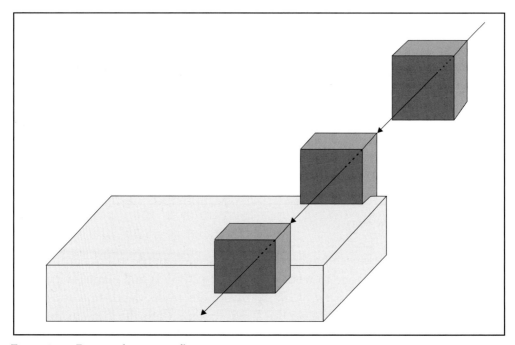

Figure 8.2   Z-minus dressing — flat.

The electrode will be rapidly redressed and should "spark out" within the first 0.050" to 0.100" of engagement, "cutting air" the rest of the way. Although "cutting air" wastes several minutes, we do this to allow for dressing block wear as the job continues through 15 more cavities.

After dressing, the electrode should stop as soon as it clears the front edge of the dressing block. Before returning to the main program, the control should reset the Z-axis to zero so it "knows" the electrode is now at least 0.010" shorter.

Switching back to coordinate system #1, the Z zero is reset by touching off a tooling ball or the parting line itself. It then reenters cavity #1 to spark-out the wear. This shouldn't take very long as there is only a 0.010" radius left in the bottom corners. When this is done, the electrode is withdrawn, leaving almost dead-sharp corners in cavity #1. It then positions itself over cavity #2 and begins rough-ing with a virtually brand-new electrode.

In this manner, the CNC EDM can run far into the night, totally unattended as it alternates from the main (cutting) program to the sub (dressing) program. The example below demonstrates this operation for a single cavity. Multicavity appli-cations would require incorporating a subprogram or macro of the dressing operation.

```
G00 G54 XYZ.1      (Position electrode over cavity)
E0000              (Load cutting parameters)
G01 Z-.5           (Machine to Z depth, return)
G00 G55 XYZ-.01    (Position electrode behind dressing block with 0.010
                    interference in Z-axis)
E0001              (Load dressing parameters)
G01 Y2.2           (Dress in Y-axis until clear of block)
G92 Z              (Re-set Z-axis)
G00 G54 XYZ.1      (Return electrode to position over cavity)
```

Real-world programming would include positioning coordinates for cavity loca-tions, computer X-Y mirrors to simplify and condense programming and a tooling ball to precisely reset the Z-axis upon returning to the main program.

MORE

We have just covered the most common and easiest dressing applications — what I call Z-minus dressing, named after its machining direction. The shape was kept simple to keep the example simple. Hopefully, you understand how this works and we can get a little more involved. Once you understand the basic concept, your imagination will steer you into more innovative dressing solutions.

We seldom enjoy an electrode shape as simple as a flat-bottomed square, so let's pretend the print calls for 45° angles on two sides, let's say 45° by 0.050".

Mill, grind, or wire-EDM this shape across the dressing block. Center the elec-trode over this shape during setup and dress off whatever wear you measured after roughing. If the chamfers are required on all four sides, simply index the electrode 90° (CNC or manually), and dress back through the block to dress the remaining two angles. (See Figure 8.3.)

This shape does not have to be symmetrical like the examples we have been using. It could have 30° angles on two opposite sides. Simply make this second electrode profile right next to the one with 45°. Dress through the 45° detail, shift the table to the 30° detail, index the electrode, and dress back through. Get the idea?

STILL MORE

Part of the reason we ground our dressing block on all sides was so we could use

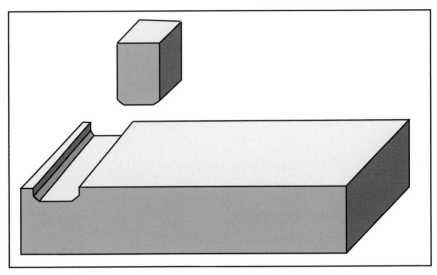

Figure 8.3   Z-minus dressing — form.

them. Let's make another electrode from our 1" square one. Let's make a rectangle 0.950" X 0.900".

Jog the electrode off to the side of the dressing block. Edge-find in the X-axis and set zero. Jog away and park the electrode behind the dressing block as in the previous example. Position the electrode into the dressing block 0.025" (minus spark gap) and proceed to dress in the Y-axis (see Figure 8.4a). When the electrode has passed completely across the end of the block, stop, index the electrode 180° (8.4b) and dress the opposite way (8.4c). Stop and measure the width of the elec-trode (8.4d). If correct, set the control to move in one-half the difference of the adjacent sides, index 90°, and dress. When this pass is finished, index 180° and return. Your rectangular electrode is finished.

Problem: Perhaps our electrode has 3° draft on each side. If we only dress off the bottom, it will quickly become too large to use. No problem.

Instead of dressing the corner wear from only the bottom of the electrode (Z-minus dressing), grind or wire-cut a 3° angle on the side of your dressing block and remove electrode wear in this manner. Either increase the electrode orbit the same amount that was removed from the sides, or dress the bottom of the electrode also. This produces a brand-new, slightly shorter electrode. Don't forget to rezero the Z-axis in both the dressing and main programs, or the machine will get "lost" on the next cavity.

ELECTRODE ROTATION

We have covered dressing electrodes in a static mode and indexing. Next we'll cover dressing while rotating, which is an incredibly useful production tool.

Almost everyone knows the advantages of EDMing with a rotating electrode, such as when machining a hole using brass or copper tubing, producing subgates

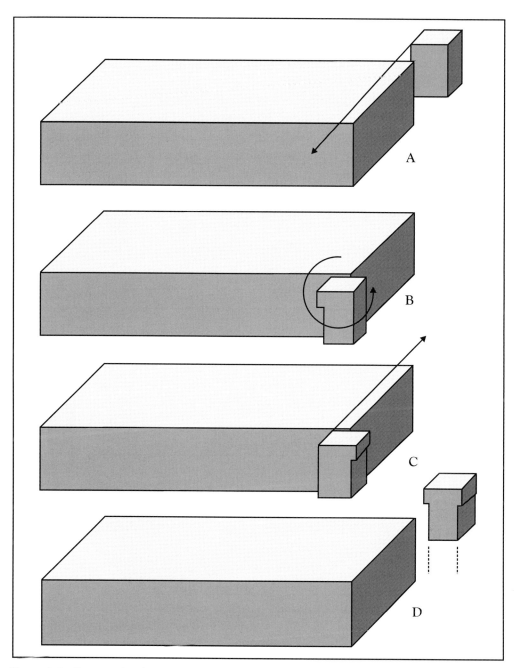

Figure 8.4   Rectangular dressing.

for molds, and other applications that require round electrodes. Rotation offers the same advantages in discharge dressing operations also. These are improved flushing and uniform wear.

Example: It is Saturday and we must "tube" a hole on a rush job. Its finished diameter must be 0.093" but, by mistake, that is the size of tubing that was ordered, with no provision for overcut!

Problem: Our tubing is too large, it's Saturday, and the supply houses are closed. What do we do?

We set up our job like we normally would, and while holding the oversized tubing in a pin chuck while rotating, traverse it across the end of our dressing block, removing just the right amount to allow for the overcut — or even a few thousandths more to allow for a "skim cut" like a wire EDM would do. Results are fast, CNC accurate, and the electrode is straighter and more concentric than it could ever be otherwise. If the hole has draft, again, no problem. Grind the correct amount of draft on the dressing block before starting.

This procedure is especially useful when the length/diameter ratio exceeds 5:1, which makes turning or grinding very difficult or downright impossible, especially in small diameters.

Discharge dressing using electrode rotation is not limited to straight or tapered electrodes. It can be used for dressing the full-radius tip and angular sides of subgate electrodes, for example. (See Figure 8.5).

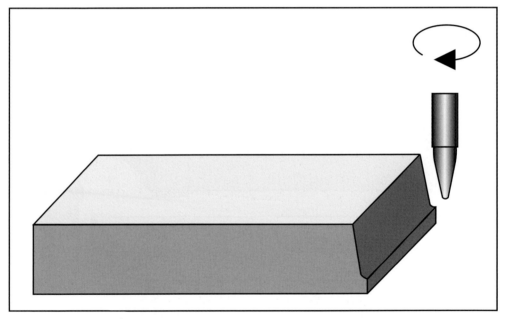

Figure 8.5   Rotation dressing.

Grind or wire-cut one-half of the electrode's profile down the side of your dressing block and set this up as described earlier.

Pass the worn electrode along this surface while rotating. Your electrode will be dressed perfectly round, accurate within a tenth and, best of all, it was done in the EDM itself, without tying up a grinder, the tooling, or the toolmaker to do it.

Some of you may ask, "Wait! Why put a radius on the end of a subgate electrode? This has always been done with a sharp electrode."

I can explain this but we'll have to call "time-out" in discharge dressing long enough to do a little "Moldmaking 101." Those of you who aren't involved in moldmaking, please bear with me — this won't take long.

SUBGATES

Subgates are a self-trimming method used to fill mold cavities with plastic. Depending on the part's shape and design and the plastic used, they can have a blueprint callout of a 0.030" diameter pin pass, or a 0.045" diameter pin-pass, etc. A gauge pin will slip-fit through the conventional subgate opening, but the opening itself is oval shaped and this shape can vary depending upon the included angle of the electrode and the centerline angle of the subgate itself. (See Figure 8.6.)

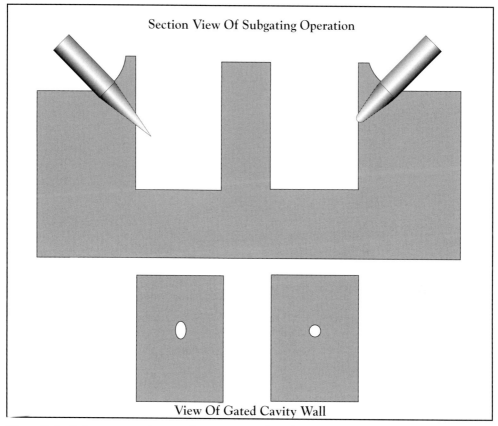

Figure 8.6   Subgate comparison — sharp versus radiused.

To EDM a perfectly round, self-sharpening subgate that leaves almost no blemish or witness mark on the plastic part, the tip of the electrode should have a radius larger than the pin-pass requirement. For example, if the pin pass is a 0.030" diameter, the tip of the electrode should have a full radius of at least 0.040" diameter. After breaking through into the cavity, allow the electrode to advance another 0.015" (allowing for spark gap). You should have a perfectly round, 0.030" diameter subgate. *Note*: This gate will restrict the flow of plastic slightly compared to the "accepted" elliptical subgate due to its smaller area (round versus elliptical), but it can easily be made slightly larger to compensate. This is a small price to pay for improved process control and better part appearance.

Bonus: The radiused electrode will EDM much better and faster than the sharply pointed one. This subgating procedure is used in manual machines also; the only difference is that the electrode is usually not redressed in a manual machine.

Sorry for the detour into moldmaking, back to discharge dressing.

CONTOURED ELECTRODES

Electrodes with shapes and contours on them can also be dressed/redressed. Many molded parts have undercuts or shapes that warrant incorporating sliding cores or cavity walls in their design. These can be cores for small electronic components to the convoluted bellows-shaped cavities required for shock absorber boots or any similarly shaped detail. (See Figure 8.7.)

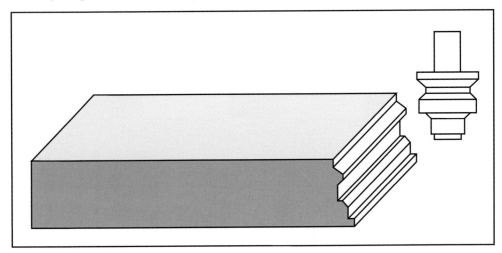

Figure 8.7   Rotation dressing — contoured.

Just as in the previous examples, the concept is the same, the only difference is in the amount of imagination used. The electrode's profile is ground or wire-cut onto the side of the dressing block and located near the cavity just as before. Since the electrode's largest diameters do the most work, they will suffer the most corner wear.

When redressing, remove only enough material to provide total cleanup. Measure this amount either by physically measuring the electrode or by using the machine's edge-finding capabilities and touching off a tooling ball or other reference surface. The difference in size from the previous electrode can be automatically entered into the program as an offset just as a wire-EDM or CNC mill would, and used on the next detail. This electrode can be dressed and redressed until its minor diameter becomes too small to reliably support the rest of the electrode. Besides proving fast and efficient, costs are reduced even further by making more complete use of raw materials.

DRESSING PARAMETERS

All of this sounds wonderful, but what kind of settings are used to do this?

Having successfully performed this operation on several different brands of EDM equipment has made me confident that this procedure can be executed on most modern CNC EDM's. Some machines will dress faster than others depending upon generator circuitry, but all should be able to perform this operation satisfactorily.

The first thing we should do is to provide positive polarity; that is, the *electrode* should carry positive polarity. Why? The dressing block should carry the negative side of the current which will provide the fastest rate of material removal.

Next, set the control to a very high frequency; that is, a very low on-/off-time. My greatest successes in discharge dressing have been with the on-time set somewhere between 5 and 10 μs.

Lower on-times can be used, but the shorter the spark duration, the less work that is accomplished. While we desire high rates of electrode wear, we also want dressing speed.

Off-time should likewise be held as low as possible to maintain efficiency. A 50% duty cycle or more can almost always be attained, because at these frequencies, the current is on for so short a duration there is very little damage to the dielectric oil, therefore it requires very little off-time to recover. A good starting point could be with a 7μs on/7μs off setting, and experiment from there — perhaps arriving at a 7μs on/2μs off setting, which is almost an 80% duty cycle.

Power or current for dressing does not seem to be as critical as it is when EDMing conventionally, as long as the area to be dressed is overpowered by at least 5 to 1. High power is strictly for speed, as finish is not important as long as we can orbit the electrode to smooth the final finish. To dress the 0.010" from the bottom of our 1" electrode in the first example, 6.5 amperes is minimum, with 10 amperes being optimum.

One of the greatest speed enhancements for discharge dressing is the addition of capacitance, with the more capacitors used, the faster the machining speeds. To establish a performance benchmark and to provide others with some idea of dressing speed, I have removed 0.100" from the bottom of a 1" square copper electrode in approximately 4 minutes. This was accomplished on a CNC-controlled EDM, using approximately 30 amps and maximum capacitance.

Of course, at this kind of speed, the dressed surface was rough; but as long as the orbit used during cavity machining is larger than the crater diameters on the electrode, this should not prove to be a problem. If finer finishes are required, rough dress on the first pass for speed, then step down all cutting parameters and make a second, finishing pass, removing only enough electrode material to change the finish.

In my experience, discharge dressing is far less sensitive to flushing conditions than normal EDMing. Perhaps this is due to the naturally occurring, larger spark gaps from the high-current settings and/or the minimal recovery time required by the oil because the on-time duration is so short. An external flushing wand or the electrode's existing through-hole flushing should provide adequate flushing for most dressing operations.

Since all makes of machines are different from one to another, and the myriad of applications are even more so, the suggestions made here are only guidelines and recommendations for a starting point. The final settings used by an operator on a certain machine on a certain application are arbitrary and entirely up to that particular operator's perception of success.

GRAPHITE

OK, what about graphite? Good question. Since 84% of all electrode material sold in the U.S. is graphite, what can be said or done for the majority group? When it comes to discharge dressing — sorry, not much. Discharge dressing graphite *will* work, but only at painfully slow erosion rates and cost-prohibitive rates of wear of the dressing block. For all of the same reasons that favor EDMing *with* graphite, discharge dressing is *against* it.

What we *can* do is offer half of a solution. While we can seldom economically discharge dress graphite electrodes, we can, however, redress certain electrodes that require rotation. Consider the following application based upon actual experience.

DRESSING GRAPHITE ELECTRODES

Several years ago, I was asked to produce a prototype mold made of aluminum. This was to quickly supply a vendor with some badly needed production parts, with the promise of a similar, Class-A mold to follow. Both molds would have 49 cavities in a 7 X 7 layout, with the cavity and electrode shape very similar to that of a subgate, only much larger, at over 3 inches in cavity depth. The procedure was as follows.

For the aluminum mold, copper electrodes were used. They would not only burn cleaner in aluminum, they also readily lent themselves to discharge dressing, so electrode wear wasn't a serious consideration. Five electrodes were prepared, with each one expected to produce at least ten cavities.

The dressing block was wire-EDMed with one-half of the cavity's profile, just like we did in the subgate example earlier. The dressing block was located next to the workpiece in the EDM machine and was set up using an unused electrode. After roughing and finishing the first cavity, there was approximately 0.030" end wear and a slight amount of taper. The control changed work-coordinates and the electrode was directed to the dressing block and was positioned 0.040" below the preset zero. This easily redressed the full radius and allowed full cleanup of the taper.

When this was accomplished, the new zero was set and the electrode returned to the main program and work-coordinate #1. A parting line touchoff was accurate enough to rezero the Z-axis and the electrode reentered the semifinished cavity to spark out the wear left by the first pass. In this manner, ten cavities were finished before changing to the next electrode. *Results:* All 49 cavities were EDMed using five copper electrodes (all had additional life remaining), and all cavities were within the ± 0.001" depth dimension. With an automatic tool changer, this job (or similar ones) could run totally unattended to completion. Even without a tool changer, this method would require operator attention for only five electrode changes.

This temporary mold was produced very quickly and took the immediate delivery pressures off the vendor. This provided the time necessary to construct the permanent, tool-steel mold that was to be used for long-run production. Using copper electrodes to EDM aluminum is fine, but tool-steel workpieces should be EDMed with graphite electrodes to take advantage of its machining speed and lower wear. The only drawback in using graphite electrodes in this application is that we will not be able to discharge dress them.

Does this mean we will have to make dozens of graphite electrodes in order to complete this job or depend upon an operator to redress them in a grinder?

The answer is no.

Remember, *with CNC EDM, we are limited only by our imagination* — let's get creative.

Before starting this job, clamp a toolmaker's vise to the EDM worktable off to the side of the mold insert. Into the vise, clamp the blotter-toolbit-blotter "sand-

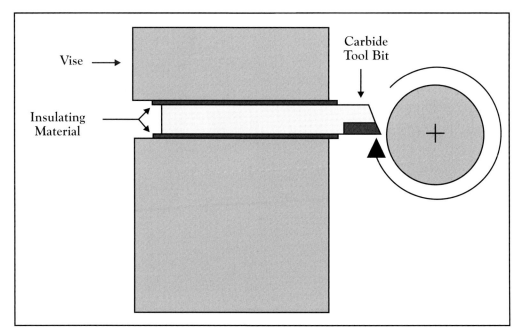

Figure 8.8   Vertical turning setup seen from above.

wich" shown in Figure 8.8. "Blotters" can be the cardboard backs from the numerous scratch pads found throughout any shop. Other (more permanent) ones can be
made from thin sheets of acrylic, fiberglass, or mycarta. This insulates the toolbit
and prevents the electrode from shorting out when it comes in
contact. (**Important!!** Before edge-finding the insulated toolbit, *bridge the toolbit to
the worktable with a cable or ground-strap* so it is "live." Check for continuity with a
multimeter. After completing the following set-up operations, *remove it.*) Manually
jog the electrode over the toolbit and in work-coordinate #2, first set the Z-axis off
the point of the toolbit, and then set the Y-axis on the electrode centerline.

Instead of an *electrical* redressing program, we will write a *mechanical* one.

For this operation, the EDM machine will be essentially a vertical lathe. Using
high-speed spindle rotation, program the electrode to face across the tip of the
toolbit, stopping right on center. Then, to regenerate the worn radius and taper,
advance the Z-axis as the X-axis moves away from center (see Figure 8.9), effectively CNC turning the electrode!

This turning operation can be done either submerged in oil or under a stream of
oil from a flushing wand. This will prevent the graphite dust from infiltrating the
air. The resulting graphite dust from machining will be captured and suspended
within the liquid oil medium and carried directly into the filtration system.

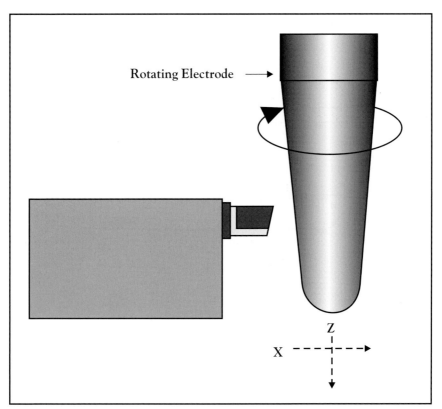

Figure 8.9   Turn-dressing graphite electrodes.

This program is carried out just like the previous copper electrode program, except instead of a "burn-dress-burn" cycle it is a "burn-turn-burn."

This method of electrode turning is applicable to almost all round graphite electrodes and is limited only by the physical or dimensional limitations of the electrode's shape, the cutting tool's geometry, and the programmer's imagination. Not bad. Many other disciplines offer much less.

Warning! *Obviously, you should not attempt to turn metallic electrodes. Besides the problem of chips and shavings falling into the cavity, the torque requirements for this probably exceed that of the rotation device. Do not attempt this operation without first checking with the builder of your rotational device whether it is integral to the machine or an aftermarket addition. It may have limitations you need to be aware of.*

ADVANTAGES

Some advantages of CNC electrode discharge dressing/fabrication are:

1) reduced raw electrode materials,
2) less electrode preparation,
3) reduced direct labor costs with unattended operations (CNC),
4) dependence upon support equipment for redressing eliminated,
5) electrodes have dead sharp edges with no burrs,
6) the ability to dress very fine or delicate details,
7) accuracy,
8) repeatability, and
9) process control and predictability.

All of these, ultimately, will reduce your EDM operational costs significantly. This, in turn, can help make you more productive, competitive, and profitable.

SUMMARY

We have just examined a few ways to use discharge dressing. The concept is exciting. Just think — if your milling machines could resharpen their own end mills whenever necessary, load a compensating offset, and resume cutting. Incredible! Even more exciting is the challenge this kind of capability offers us — the users, designers, planners, and engineers.

Today's modern EDM equipment is capable of so much, but it's up to *us* to figure out new and innovative ways to use them, to exploit them, and to profit from them.

Discharge dressing is one of them.

Chapter 9

PREVENTIVE MAINTENANCE

The following recommendations are generic guidelines prescribed to maintain and keep your vertical EDM equipment functional and profitable.

FILTRATION SYSTEM

FILTERS
Filter change intervals are to be determined by the end user, depending upon frequency and degree of use. The pressure gauge (on the dielectric reservoir) is indicative of filter life. On many newer models, the color graduated dials are self-explanatory. On older models, consult manuals and change filters when the needle reaches or exceeds the pressure specified by the manufacturer. If manuals are unavailable, a telephone call to the manufacturer can usually obtain the information. Make sure to vent the filter canister(s) with the pump on to purge air trapped in the system during filter change.

Recommendation: Keep at least two sets of filters in-house as requisitions and/or deliveries are not always made in a timely manner.

WORK TANK
The operator should check the work tank door seal daily. It should be firm and resilient to the touch and free from leaks.

Recommendation: It is advisable to keep several lengths of replacement seal material in reserve, as this is a consumable item with no predetermined life span.

On machines equipped with rise and fall tanks, check the appropriate rubber seals and bellows. Make sure that the screw and nut actuators are lubricated properly.

Keep the inside of the work tank clean. A black carbon film will adhere to the inside surfaces of the work tank, decreasing the amount of reflected light which will make the parts and setup harder to see and is, in general, aesthetically unappealing. This film can be easily removed at the end of every workday with an inexpensive 2" or 4" wide paint brush while the tank is still wet with oil. "Sweep" any slurry or debris into the return drain. Turning the pump on and "hosing down" the inside of the tank will make the process easier.

On work tanks with a painted surface, a soapy spray cleaner can be used periodically, but avoid the use of harsh solvents that can strip the "glaze" from the finish, making subsequent cleanings more difficult.

RESERVOIR

Inspect monthly. Do this on a Monday morning before turning the pump on, because the oil will have had at least one day for the sediments to "settle out." Do this visually with a flashlight, or use a wooden "dip stick" to gauge the depth of the sediment. Remove sediment and debris from the bottom of the dielectric reservoir periodically. Unless under severe use, once a year should prove adequate.

DIELECTRIC OIL

Depending on its use and maintenance, dielectric oils can last for several years. Regularly scheduled filter changes and prevention of water contamination will extend its useful life considerably.

Water contamination cannot be avoided entirely as condensation will occur on the electrode surfaces as it heats up. Graphite electrodes will contribute more to this condition than metallic electrodes, as their porous structure will actually attract and absorb moisture from the air. Storing graphite electrodes in a clean, dry area will keep moisture absorption to a minimum. Some shops using very large graphite electrodes will place them in a low-temp drying oven the night before their intended use (but this is an exception, not the rule). A less obtrusive method to keep humidity and moisture absorption to a minimum would be to allow a 60 watt lightbulb to remain lit within the storage cabinet.

The color of the oil is not necessarily an indication for oil replacement. All oils, no matter how clear they are when new, will darken in shades from amber to brown with use and age, because these products will break down when exposed to high heat. Obviously, sustained high amperage machining will break the oil down more rapidly. Tars, resins, and hydrocarbons are generated when this occurs, and this is what "stains" the oil. No amount of filtration will remove this discoloration, so don't mistake "colored oil" for "dirty oil."

Dirty oil is best judged by several factors.

1) Pressure gauge readings as described in the machine's maintenance manual.
2) Increased occurrence of dc arcing or pitting with settings that were previously successful. (Assuming, of course, that no other changes have been made, such as the grade of the electrode material, flushing pressures, etc.)
3) Longer cutting cycles and/or degradation of finishes.
4) A visual inspection of the oil. To do this, fill the work tank but do not machine. Collect a sample of oil in a clear container. Visually check the sample immediately for any cloudiness, and again after a few hours to check for sediments on the bottom or color striations in the oil itself. Dirty oil is usually tinted gray or black, and this coloration will dissipate if the oil is allowed to circulate through the filtration system. Replacing the filters is the least expensive and the most likely remedy to the above symptoms.

If a filter change reduces pressure gauge readings but does not correct the deteriorating cutting conditions, then replacement of the oil may also be warranted. Generally speaking, unless the machine uses only very low power settings and/or high frequencies, it is safe to consider replacement of the oil after 18 months to 3 years of daily use.

Replacing your EDM oil can be an expensive proposition, especially in larger machines and central filtration systems. To be absolutely sure when this is necessary, there are several firms that can test a sample of your oil for acidity, hydrocarbons, water, oxidation, and other contaminants, and they can advise you in a much more scientific manner than I have been able to do in this chapter.

In the event an oil change is required, drain the reservoir completely, and remove all sludge and sediment. Wipe the greasy film from the sides of the sump and inside of filter canisters before refilling with oil and replacing filter elements. (*Note*: Check with local authorities concerning the correct methods of waste oil and filter disposal.)

If the dielectric system is exceptionally dirty or has not been used for several months, partially fill the sump with just enough fresh oil to prevent the pump from losing prime, and allow it to circulate for approximately one hour to dissolve and wash away any films or coatings remaining inside pipes and hoses. Most EDM oils contain some active solvents (especially when fresh) and will help clean the inaccessible areas of the dielectric system. Drain this oil and dispose of it properly.

Refill the reservoir with a high-quality EDM oil made especially for this purpose. Some shops use deodorized kerosene with success, but keep in mind that this is not a product that has been engineered specifically for EDM use. Those who use it like its low viscosity and finishing capabilities, but it has many undesirable qualities such as low flash point, high evaporation rate, and it will cause excessive defatting of the skin, resulting in irritation, redness, and possibly dermatitis.

Caution: Due to its low flash point, deodorized kerosene may not meet city or state fire codes and other regulations in your area. If you are already using it, check with your local authorities concerning its safety and legality. If you are considering using it, don't.

FIRE PREVENTION SYSTEM

FLOAT SWITCH
Check the float switch mechanism weekly by attempting to machine with the pump on, but with the waste gate open. The float switch should not allow machining if the dielectric level is low. This device is frequently disabled by operators for various reasons, but the machine should never be allowed to run unattended without it.

FIRE EXTINGUISHER
Check the fire extinguisher and related systems once a month. This includes a battery check and system charge verification using the test buttons and/or pressure gauge on the fire extinguisher. In addition, ensure that the detection sensors around the head are not bent or damaged and that the nozzle outlets are unobstructed. Consult your machine's manual for information concerning recharging.

MACHINE TOOL

LUBRICATION
Grease machine periodically as per maintenance manual, using specified lubricants.

POWER CONNECTION BRUSHES
On rotary spindle machines, periodically inspect for excessive dust or vapor contamination. Clean with contact cleaner when necessary.

AUTOMATIC TOOL CHANGER
Drain the water separator reservoir daily and check for adequate air pressure. Remove the cover and inspect fluid levels semiannually. Lube visible grease fittings with the grade specified by the manufacturer.

CONTROL/POWER SUPPLY

POWER CABLES
Inspect power supply cables quarterly. Check for clean, secure connections at the machine tool and at power supply terminals. Power cables should be free from breaks or kinks.

CABINET

Visually inspect all sealing surfaces of cabinet panels upon their removal to ensure an air-tight seal. There should be an "imprint" pressed into the rubber seals indicating a good seal when the mounting screws are tightened. The intake fans are powerful and should draw air from only the intake screens.

Ensure that exhaust ports are clean and free from obstructions. This will ensure adequate air circulation.

INTAKE FILTERS

Inspect and clean all intake filters periodically. This interval should be determined by operators and maintenance personnel depending on shop conditions.

Indications of more frequent attention would be if the shop environment involves other machine tools that are utilizing coolants and/or cutting fluids that can result in their vapors being drawn into these intakes. Also, grinding operations that can suspend metallic particles in the air are especially hazardous to electronic circuitry. If these conditions cannot be prevented or controlled, frequent attention to this area is highly recommended.

CIRCUIT BOARDS

Most shop environments offer reasonable control over airborne vapors and particles. In these conditions, a yearly check of circuit boards is sufficient. In areas where coolant mists and/or grinding dust is prevalent, more frequent inspection and cleaning is required.

If necessary, carefully remove circuit boards *one at a time* to avoid any mixups, and gently clean any dust or "fuzz" from the circuitry with a clean soft brush.

Caution! Do not use an air hose on the boards or within the cabinet!

Before reinstalling the cleaned boards, clean the contacts with a contact cleaner designed for this purpose. *Do not* rub them with a pencil eraser or with Scotchbrite. This can damage the fine plating and actually defeat the purpose of achieving good contact.

DISK DRIVES

To prevent data loss and distortion on CNC machines with disk drives, it is recommended that each drive be cleaned with disk cleaner recommended or supplied by the builder. Commercially available cleaners should work fine also, but check for compatibility to be sure. This should be done at least once per month, and more often if frequently loading or downloading programs or running directly from disk.

SOFTWARE

Make frequent backups of program disks. Besides the occasional rare disk failure and the resulting loss of information, fluctuations of incoming power while editing or saving data can scramble or erase valuable programs. If this is a common occurrence (more than twice a year), and is not attributed to disk failure, it is strongly suggested that obtaining voltage suppression may be warranted. (This is protection that transformers, voltage regulators, or a "dedicated line" cannot provide.)

Duplicate *all* system disks and store them in a secure area. This will safeguard against job program loss as well as actual EDM system data loss and perhaps the resulting inability to use the machine. EDM system data can consist of cutting conditions, offsets, pitch compensation, etc. This type of information is integral to the operation of the machine, and you should always have backup data available. Without a backup disk, one would not only need the latest printout of this information, but all of these data would have to be manually reentered into memory. This amount of data entry would be very tedious, time-consuming, and would carry a high probability of error.

Three copies of the machine's system data or "brains" should be made: one for the machine's daily use, an operator's backup copy stored in a compartment of the machine or a nearby workbench drawer, and a third copy to be stored in a secure area away from the machine in an office or where the machine's records are stored. Multiple backups provide good insurance against the loss or damage of the originals.

GENERAL

DOCUMENTATION

All records, service reports, internal memos, etc., pertaining to the machine should be copied several times and stored: 1) in or near the machine itself; 2) in the Maintenance Department file; and 3) wherever the original purchase order agreement is filed.

Redundancy in this area is to ensure against loss or destruction. This is to document the machine's history from the time of installation. This can provide important data that can assist in troubleshooting by indicating possible trends in machine performance, or assist in the future trade-in or resale of the machine.

If you are lacking any book, manual, or instructions for a machine, call your distributor or machine manufacturer. If your machine is older or no longer being made, contact someone in your area with the same equipment to see if they will give or sell you copies of their documents.

FACTORY SERVICE AND SUPPORT

If at any time there is a question or problem pertaining to any aspect of your machine's operation, performance, or maintenance, do not hesitate to contact your local distributor or manufacturer. Keep telephone numbers of local service and applications engineers near the machine for easy operator access.

SUMMARY

These procedures are only guidelines, considering the numerous makes and models of equipment in the field. Therefore, these instructions have been kept very generic for obvious reasons. Consult the manufacturer's manual for specific instructions whenever possible.

Whatever maintenance program is ultimately set up for your equipment, let it be exactly that — *a program* — a regularly scheduled program of maintenance, with some items being tended to or monitored every day, and other items that require attention annually. The key word in "Preventive Maintenance" is "Preventive," as in *prevention*. A small amount of downtime for regularly scheduled preventive maintenance is far less costly than the expense of something major occurring that could have been avoided by regular maintenance or early detection.

If you feel that the time required for preventive maintenance is too costly, how much have you budgeted for downtime?

SECTION III

WIRE EDM

Chapter 10

PREPARING FOR WIRE EDM OPERATIONS

INTRODUCTION

Things are not always quite as simple as they may seem. Whether you have purchased capital equipment before, or you are considering your first venture into wire EDM, you will find its integration totally unlike that of any other type of equipment. There are many things to consider — things above and beyond the actual machine selection.

Some of the decisions involved won't be easy and therefore, shouldn't be taken lightly. This book certainly isn't intended to frighten or intimidate anyone considering wire EDM as a business venture, after all, it was written to *promote* the EDM process. But since there have been so many requests to provide this kind of information, I would be remiss if I didn't point out the many different and sometimes unforeseen facets involved in the successful initiation and integration of this process. Are you ready?

THE WORK ENVIRONMENT BEFORE INSTALLATION

After deciding upon the make and model of wire machine that will fit your needs (see Chapter 20 on "How to Select and Justify an EDM Machine"), check with the builder of your EDM for its electrical, water and air requirements. Have these items prepared and ready *before* the machine is delivered. — it is always easier to install these items before the machine is in place.

Next, you must allow room for the actual installation. Forklifts and riggers need sufficient room to maneuver heavy equipment into place. This planning includes not just the machine's work area but also the height and width clearances through doorways or access ways *en route* to the actual placement site.

The placement of the machine is very important. The success or failure of a wire EDM operation can depend upon its environment. There are many things to consider before the actual purchase, starting with: How much space will the machine require? Where will the machine be placed? (We will take for granted that the machine will be installed on a solid foundation.) Will it have its own stall or room? Will it share a room with other EDM's? Will it share an area with chip-making machinery?

Obviously, you should never install a machine in direct or partial sunlight. This can cause temperature variations of different components of the machine that can affect part size and consistency. Likewise, the machine should be protected from drafts caused by open doors or windows. Heater and air-conditioning ducts should be directed away from the machine also.

It is usually best when a wire machine has its own area or shares an area with other wire machines. This way the working conditions for multiple machines will all be the same so the parts they will produce will be more consistent and easier to control.

There should be provisions for filling and draining the dielectric reservoir. There should be a water tap installed at the machine, or in close enough proximity to use a garden hose to fill the tank. Many machines have rollers or casters on the dielectric reservoirs that allow them to be pushed outside for draining. A floor drain in the room itself is ideal because not only does this allow draining of the tank without moving it outside, but it is also good to have in the event of an accidental spill (this is no longer an experience unique to vertical EDM as more submerged wire-cut machines are put in use).

Note: Do not dispose of any waste water (or filters or resin for that matter) from your wire EDM equipment without first checking with your local, state, and federal authorities on the correct methods and procedures of doing this.

While the recent trend of builders has been the downsizing of the equipment to have a smaller footprint and save valuable floor space, be sure to allow sufficient access on all sides of the machine. There will be routine preventive maintenance to be performed, as well as filter and resin changes, etc. Don't forget to allow for the not-so-routine maintenance also. Situations like pump replacements require adequate workspace to lift and position sometimes heavy and awkward pieces of equipment.

One final note: EDM's emit a great deal of electrical radiation from the power cables and the wire electrode. These are in the forms of high-frequency radio, infrared, and electromagnetic radiation and noise. These emissions are harmless but can affect computers, televisions, AM radio or other electrically sensitive devices that are in close proximity. In certain cases, line filters and/or electric shielding may be required. Space does not allow for elaboration here, so if you feel your work environment is electrically sensitive, consult an electrical engineer for recommendations.

INCOMING POWER

Power quality is one of the most overlooked aspects of any CNC operation, and CNC EDM in particular. This type of equipment, especially the computer section, is typically very sensitive to fluctuating or "dirty" power. There have been numerous occasions where customers have had strange incidences occur with their EDM equipment, regardless of its make, model, or age. Unexplained CRT scrambling, CPU lock-up, memory dump, the machine "zigging" instead of "zagging," etc., are all indicative of power problems.

To the layman, who is usually unaware of the symptoms that a power problem can display, or the consequences suffered from the lack of power protection, your first step is to call a service engineer to fix your machine. Quite often, when the technician is on site, the problem or condition cannot be duplicated, and both of you end up wasting considerable valuable time.

The technician is well trained and professional, but finding nothing presently wrong with the machine, he can logically conclude that the problem was probably "operator error." The owner or manager, upon being told that the service engineer could find nothing wrong, could easily conclude that the engineer is either inexperienced or inept. After all, he knows for a fact that the machine inexplicably scrapped his part. It is often very difficult for an owner or operator to accept the fact that there may be "ghosts" in the machine, but sometimes there are.

There are several facets to power problems, and this can be very complicated to the layman, but we will try to keep this simple as we start from the top.

POWER COMPANIES

Electrical power is generated and distributed by the power companies. They are required by the Public Utilities Commission to deliver power to your terminal but they are not required to guarantee the quality of that power beyond a tolerance that is typically ± 5 to 10%.

Then, after it leaves your plant's main distribution panel but before it reaches your EDM machine, it can deteriorate another 5% as air-conditioners, welding equipment, and even office laser printers make their demands upon it.

In many states, from the supplier's point of view, a voltage "problem" doesn't exist unless there is an occurrence or disturbance that lasts for 30 cycles or more. Since U.S. power runs at 60 cycles per second, according to these parameters, the power must be "bad" for one-half second before warranting attention by the supplier.

Wait just a minute. A half a second? Why, some of the computers in modern EDM equipment are running in excess of 25 million cycles per second (25.4 MHz). So, in a half second, over 12 million cycles — right or wrong, clean or dirty —

will pulse through your computer and power supply. This means that the supplier can deliver power that can be bad for almost 12,700,000 computer cycles and still be considered "good." Maybe so, but not good enough.

The following devices are used to prevent some of the problems caused by poor power.

TRANSFORMERS

These are the most often used means of power regulation, although they are mistakenly given far too much credit for solving power problems. A transformer merely steps the power up or down to the machine's recommended voltage. It has no control whatsoever over transients, spikes, surges, sags, or noise. Unfortunately, most EDM installations can require much more, but this is a good start. Transformers for a typical wire EDM installation are priced anywhere from $500 to $1,000.

DEDICATED LINES

Often, manufacturers specify the requirement of a "dedicated line." This is a source of power to the machine with no other equipment or taps on it, although the viability of this concept is misleading. While this can help in power management, it is by no means a guarantee of trouble-free power.

For example: you know that the incoming power to your operation is 220 V ac. At least, that is what it is supposed to be. But during the day, when all of your manufacturing neighbors are busy, including air-conditioning during peak usage in the summer, power may sag to 200 or even 190 V. In the evening, when it is cooler and many shops have shut down, power can be slightly higher, perhaps 215 V.

Naturally, your wire machine is going to behave differently with these kinds of fluctuations. This alone can be the reason it seems to cut faster on second shift than it does during the day. Speeds, feeds, computer response time, etc., can all be affected. Incoming power can be equated to octane ratings for gasoline. You know very well that when using low or high octane fuel, most vehicle's performance will drop off or increase accordingly.

GROUNDING

All EDM equipment must be properly grounded, but conduit grounding is not adequate for the computer-side of the power supply. That is, grounding to the conduit pipe itself is not sufficient (much different than the copper ground-rod used in most EDM installations). Proper grounding is even more critical when connected to other machines or computers by communication or interface cables. All units in this circuit must have the same ground reference or the differences in electrical potential can cause computer errors.

VOLTAGE REGULATORS

Keep in mind that the power we use is generated according to demand. The only problem with this is that the power companies can only react to conditions that already exist.

Scenario: Between 6:00 and 8:00 in the morning, office lights, copiers, computers, and other office appliances are turned on, and factory machinery is already in full swing. As the day progresses, air-conditioners are turned on, increasing demand as delivered power decreases, perhaps to the day's lowest. The power company reacts to this by bringing more power on-line.

About this time, the whole city breaks for lunch. Power surges as machinery and heavy equipment are turned off. Perhaps the power company lowers its output to compensate, but at 1:00 PM, all of the electrical machinery is switched back on. The power company must respond once again.

By 3:00 PM, demand starts to decrease as plants and factories start shutting down for the day, but power generators are still running at full-tilt. By 5:00 in the afternoon, power is perhaps at its highest, but actual demand continues to decrease.

This whole scenario reminds me of trying to pull the slack out of a long freight train — the locomotive demands that the cars move or stop, but they react long after the engine moves or stops. Just like the demands on power, it has almost a whiplash effect.

We can do nothing for the long freight train, but we *can* regulate the fluctuating power with Automatic Voltage Regulators (AVR's).

Voltage regulators are required whenever primary voltages vary more than 10%. An AVR will regulate the incoming voltage, and do a good job of it, usually less than ± 5%, as typically specified by many builders. Unfortunately, it does this at only 60 Hz (or 60 cycles per second). It will guard against normal sags and surges but transients can come blazing through the line in mere nanoseconds, (that is billionths of a second — a "transient" typically has a duration ranging from 10 nanoseconds to 100 microseconds). At these speeds, the voltage regulator can't even "see" the spike, let alone regulate it. The only solution for this problem is voltage suppression.

VOLTAGE SUPPRESSION

There is another type of power problem that is potentially far more damaging to sensitive electrical equipment. They are called transient spikes.

These are incredibly high bursts of current (as high as 6,000 V), sometimes caused by the power companies themselves when adjusting power or switching grids. These surges and spikes last for only a few nanoseconds but typically head straight for the most sensitive and vulnerable part of the computer — the Central Processing Unit (CPU) — modifying or destroying information. It can zip through the circuitry so fast it doesn't have time to heat up and burn through a fuse or trip a breaker. It can change a code or trip a switch, leaving no evidence that it was ever there — except maybe a jumbled screen or worse, a scrap part. How can this happen and why is CNC equipment so susceptible?

Within any CNC program, every character, number, decimal, or sign is called a byte. Each byte is comprised of 8 bits of binary coded information (a series of 1's or 0's). Changing only 1 bit of information can completely change a byte, perhaps changing a "plus" into a "minus" or an "X" to a "Y," or maybe moving the decimal point several places to the left or right, perhaps resulting in your expensive workpiece being cut in half!

Transient Voltage Protection Equipment (TVPE) consists of a device that is installed between the transformer and the control/generator at line voltage, and another that is placed inside the control for the 100 V computer. TVPE, in concert with a transformer and an AVR, can solve most, if not all of your EDM problems related to power. TVPE is relatively inexpensive, priced around $1,200 to $2,000, installed. When comparing this amount to the amount of your initial investment in the wire machine (or any CNC machine for that matter), this is very inexpensive insurance — much less than even the sales tax on the purchase.

If you don't safeguard your EDM equipment from power problems and instead use an excuse like "I didn't really budget for this," then please answer another question: "How much did you budget for downtime?"

THE WORK ENVIRONMENT AFTER INSTALLATION

So now the machine is installed in a stable or temperature-controlled environment, and the incoming power is clean and regulated. What next? Well, it really depends upon how much the machine is used and the type of use your machine will encounter.

First of all, most wire-cut machines are equipped with a dielectric chiller. This is essentially a refrigerator without a door that monitors and maintains a preset water temperature. Chillers are necessary — more because of pump operating temperatures than actual cutting temperatures.

Basically a chiller is nothing more than a heat exchanger, transferring calories from the water and expelling them from an exhaust opening; therefore, it can put a great deal of heat out into the work environment. If your work area isn't temperature-controlled, it can get downright unbearable for the operators, especially if it is enclosed.

If the work environment *is* temperature-controlled, then we can encounter a "Catch 22" situation. In the past, it has been recommended to direct the hot exhaust from a chiller through ducting up through the ceiling or out through the wall, because if this heat is allowed to return into the room, it can increase the requirements of your existing air-conditioning system, forcing it to work much harder than necessary. The logic of this seemed obvious: don't force your EDM chiller to fight with your air-conditioning.

But recently a finer point on this subject has been brought up. These units are typically very efficient, and that may very well be the problem. If the volume of room air drawn in and expelled by the chiller is more than the output of the room's air-conditioner, it will actually pump the expensive cool air right out of the room!

To remedy this situation, the ducting should be routed to *both* the intake and exhaust ports of the chiller and then routed outside the EDM room. This should allow the room to have a positive pressure instead of negative, and any conflict with the air-conditioning system is eliminated.

Another problem encountered in a closed work environment is humidity. Most installations will not be a problem as long as the relative humidity is 35 to 75%. But with the advent of high-pressure flushing, humidity can become a problem in small enclosed work areas. The mist or spray generated by high-pressure flushing can be considerable and, in extreme cases, can cause rusting of carbon steel parts in the room or cause malfunctions of computer equipment. Generally, these conditions occur only during high-speed cutting and roughing operations or where multiple machines are operating at the same time. This will not only disperse a great deal of moisture into the air, this moisture can also contain tiny particles of metal and metal oxides from the wire and workpiece. In severe cases, the air can actually have an unpleasant, metallic taste, almost like sucking on a penny! Obviously, under no circumstances should condensation be allowed to form outside the immediate work area. Imagine what can happen to the sensitive electrical circuitry if air this moist were allowed to enter the control or power supply. While situations this severe are rare, there are various aftermarket dehumidifiers and air deionizers to help solve this problem should it occur.

TOOLING

Hopefully, you have invested in a good tooling system or are planning to do so; therefore, I'm not going to spend a lot of time on this subject. I would like to think that if you are seriously considering wire EDM as an investment in business, and especially if you have already done so, you already understand the value and importance of good, quality tooling.

Unless you must use special tooling that is dedicated to long-run production, or have a job that is held in special fixtures that are commercially unavailable, your wire EDM operation needs a good balanced tooling package. If you are a first-time buyer and are unsure of the tooling you will require, consult the applications engineers of the builder of your machine or consult the tooling company directly. These people are experts and can reliably assist you in the selection of your EDM tooling.

Although all quality tooling packages may seem expensive, you must consider them an *investment*, not an expense. A complete tooling system for wire EDM should allow an operator to make setups or inspect parts accurately, away from the machine. Wire EDM equipment is very expensive and should be kept running as much as possible. Having a machine tool stand idle while making setups in the machine is not cost-effective.

Be sensible. You are investing in capital equipment possibly costing several hundred thousand dollars, so don't handicap yourself before you start. Invest in good tooling.

Figure 10.1   Presetting station for wire EDM. (Courtesy of System 3R, USA, Inc.)

PROGRAMMING

Now that we have the room, clean power, machine, and tooling, we must be able to "talk" to the machine to get it to perform the work. Most EDM programs today are far too complex to simply input manually through the keyboard (MDI or Manual Data Input), so a programming system is a must.

Some machines come with "on-board" programming. This type is handy if you don't have a CNC programmer or programming department, and your programming is to be done at the machine by the operator. For many shops, this system proves adequate, but often, these types are limited in their abilities. And if complex or multi-axes programs are required, such as involutes or curve-fitting, optional software or upgrades are required.

If you already have a programming department or wish to generate programs from a PC station in an office, then a remote programming system or station must be considered. There are many types of software available, and if you already have a programming system for machining or turning centers, then usually all that is required is the addition of an EDM "module" and postprocessor to your existing system. The programs can be downloaded to the wire machine via RS-232C, floppy disk, or punched-paper tape.

Programming software can cost as little as $1,500 and run as high as $10,000, depending upon the capabilities desired. Some high-end turn-key systems, complete with hardware, software, postprocessor, and complete integration with the machine, can run over $30,000. Once again, examine your requirements and consult with application and software engineers.

If you are an entry-level user, keep the following suggestion in mind. Right now you don't know exactly *what* you need, but odds are, as you gain experience and confidence with your new machine tool, your capabilities and skill levels will increase substantially. The scope and spectrum of work you will seek and pursue will broaden and expand, especially if you are a job shop. As far as programming capabilities go — personally, I would rather have too much than not enough.

TRAINING

Operator training is standard with any wire EDM purchase. This training should encompass: machine tool operation, basic application recognition and tooling strategies, part programming, preventive maintenance, and basic set-up techniques.

Since most wire-cut machines are a considerable expense, train as many people as the purchase agreement will allow. Having several operators will prevent illness or injury from stalling your EDM operations.

Protect your EDM investment by carefully selecting those who will be trained. It doesn't matter whether you choose your foreman or apprentice, they should be the sharpest and most ambitious of your toolroom personnel who wish to learn and grow and better their skills. Good shop experience is a plus, but is not absolutely necessary — with third or fourth-year apprentices proving to be excellent candidates. Further, your choice should obviously be company-oriented people — ones who are likely to remain your employees for many years. In other words, don't train people for your competition.

Companies with newly purchased EDM equipment often make the mistake of sending slightly marginal people for training because critical shop operations might suffer while the "good ones" were away. But the future of your entire wire EDM investment can suffer if you don't plan for the long haul.

Except in the smallest of shops, the several days spent in initial operator training shouldn't affect your business *that* much. Plan ahead by making schedule allowances *before* your EDM purchase. It can typically take several weeks for the selection, financing, and delivery of a wire machine, which is plenty of time to allow for this all-important part of your investment — operator training. Make this small sacrifice in schedules and train your best people.

The training itself can be conducted on- or off-site, depending upon the terms of your purchase agreement. Controlled, formal classroom training proves the best, but if training is to be conducted at your facility, be fair to both the students and the instructor by providing a quiet environment and uninterrupted training periods. It can be very difficult to resist the temptation to interrupt a class to ask your foreman a pressing question or to help solve a problem in the shop, but don't. Try to schedule your meetings and troubleshooting sessions around these training periods by utilizing break periods and lunches. Keep interruptions to an absolute minimum.

Training for wire EDM is valuable, comprehensive, and intense. Treat it as such.

OPERATING COSTS

At first, I felt this section could be left out because there are so many variables from machine to machine and because every shop's operation is different. But on second thought, a new or potential user should have at least an *idea* of how much consumable costs can be.

Remember, these prices are for reference only. They will vary depending on the make of machine, the quantity of items ordered, and the geographical location.

WIRE

These costs can vary depending upon the wire diameter and quantities purchased. Brass wires typically cost between $3.75 and $6.50 a pound (again, depending upon volume). Coated or stratified wires can cost as much as three times more. Fine wires, such as molybdenum, can cost from $75 to $125 a pound.

Actual wire consumption will be too complicated to calculate here, since every operation or application is different. Contact the Senior Applications Engineer of your machine builder. Acquaint him or her with your primary intentions concerning the type of work you expect to pursue, the workpiece material, thickness, finish, complexity, the number of hours of operation per day, etc. From these figures, the two of you should be able to predict reasonable estimates of expected wire consumption.

FILTERS

Filter life is also hard to predict. Workpiece materials can affect filter life significantly, with aluminums, nickel-based alloys, and graphites consuming filters faster than normal. Paper canister filters run, depending upon size, from $16 to $85 each.

If you are using or considering an after-market filtration system or central filtration unit, consult with representatives of its maker for reliable estimates of filter life.

RESIN

Mixed-bed resin is used for controlling the chemical cleanliness of the dielectric, and runs about $175 to $200 a cubic foot for virgin resin. Regenerated resins have approximately 85% of the efficiency of virgin resins at reduced costs. Again, your application and frequency of use will determine its useful life.

POWER-FEED CONTACTS

These are typically made of carbide and are used to transfer current to the wire. The serviceable life of these contacts varies from machine to machine and the degree of use. Cutting with high-power settings that are required in speed-cutting or tall parts will consume carbides much faster than cutting thin workpieces or executing fine skim cuts. Carbide contacts can run anywhere from $40 to $180 each.

WIRE GUIDES

Sapphire and diamond wire guides come in different configurations — round or split-vee — depending upon the machine. These can last between 6 months to a year depending upon use. Steep tapering and complex shapes tend to shorten guide life. Replacement costs can run between $350 and $1,400 each.

PREVENTIVE MAINTENANCE

Wire-cut EDM's can be incredibly productive and profitable machine tools; and with today's modern technology they can routinely do what used to be considered impossible. But all this capability does come with a price — a small price, but one that is absolutely required. This price is maintenance and preventive maintenance. Because of the tendency to run long hours unattended, we can often take for granted just how many hours the machine actually runs. For this reason, making and following a written timetable of maintenance procedures is often a necessity. You should also consult the more detailed wire-EDM preventive maintenance procedures outlined in Chapter 16.

Due to the many makes and models of machines on the market today, a list of only general procedures will be recommended, with cleanliness being universal to all makes.

DIELECTRIC SYSTEM

Dielectric Level. Check level frequently, especially when making high-speed cuts and using high-pressure flushing, which can consume large quantities of water. Most tanks or reservoirs have a float level indicator or a clear viewing tube along the side.

Do not allow your pumps to run dry. Besides the nuisance of stopping to prime them, it can also shorten their life.

Filters. Regular filter changes should go without saying. If you are cutting the same type of material daily or running production, you can easily determine filter life and schedule changes accordingly. If you have equipped your machine with an after-market unit or central filtration system, consult the manufacturer.

Resin. Mixed-bed resins control the chemical cleanliness of the dielectric. Careful monitoring and changing of the filters will prolong the life of the resin.

The resin itself is in the form of small plastic beads that have approximately 65% of their ion exchange surfaces on the inside. Unfiltered water can clog these very small passages and significantly shorten the useful life of the resin.

Conductivity Sensor. This is a small sensing device with two small electrodes immersed in the deionized water. A low current is passed across these probes to measure and indicate the water's conductivity. Depending upon the water quality, mineral deposits or corrosion can influence sensor readings. Check and clean periodically.

WIRE-FEED MECHANISM

Wire Spool. This may sound silly, but many times operators plan on a job to run unattended or overnight but fail to provide enough wire to do so. Make sure that the wire spool has sufficient wire to run all night. Entry-level operators are advised to load a fresh spool for overnight operations. When experience alone isn't enough to judge this accurately, calculate the remaining amount of wire of a partially used spool by its weight and the wire speed. Later, with experience, a mere visual inspection can provide a reasonable estimate of wire remaining. The chart in Figure 16.1 will help provide an estimate of the wire time remaining on a spool.

Wire Basket/Take-Up Spool. This is actually the second half of providing sufficient wire to complete the job — do not let the wire basket get too full. This can wind snags in the pinch rollers or belts. Besides stopping the machine, this can create a costly repair bill. Likewise, on machines so equipped, make sure the take-up spool doesn't become wound with too much wire. On some models this can actually cause jamming of the spool.

On jobs with a known wire consumption, this can be calculated into a maximum safe time between wire removal. This may be annoying — returning during the night or weekend — but it beats having a jammed pulley and a possibly delayed delivery of the job.

Pinch Rollers, Bobbins, and Belts. All of the rollers, bushings, bobbins, and tensioning devices in the drive train can fall into this category. All of these should be cleaned on a regular schedule, and even more often when coated or stratified wires are used. Check for grooves or flat spots on wear surfaces and worn bearings on metal parts, and worn, cracked, or frayed areas of drive belts.

Power-Feed Contacts/Energizer Pins. Inspect these often, sometimes daily, depending on use. Index or rotate them at regular intervals or sooner, depending upon the nature of the cutting, to obtain maximum speed and efficiency. Erratic cutting behavior or "brassing" of the workpiece are signs that the contacts need attention.

Wire Guides. Inspect these weekly, or more often when using coated wires or cutting tapered parts. Remove brass deposits or zinc build-up by submerging the guide assembly in an inexpensive ultrasonic cleaner. On split guides, make sure mating surfaces are clean and free of wire shavings or debris.

Flush Cups/Nozzles. Inspect these daily. Check for cracks, fractures, or any damage, particularly around the flat, "sealing" surfaces. These areas are critical to efficient flushing and cutting speeds. A cracked or chipped nozzle can reduce cutting speeds up to 30% and increase the possibilities of wire breakage. "Flushing, flushing, and flushing" is just as critical to wire EDM as it is to vertical EDM.

AUTOMATIC WIRE THREADERS

Water Flow. The flow of water leaving the upper nozzle can be visually checked for a smooth, uniform flow that is centered in the opening. Any spray, visible turbulence, or deflection of flow from center is indicative of a problem.

Guide Pipes. The high-pressure jet of water during rethreading is meant to center the wire in the pipe for a more accurate try at the start hole below. These need to be clean and free from any bends or kinks.

Parting Mechanism When moving from part to part in production (or detail to detail in die work), the wire machine must sever the wire to enable it to pass on to the next threading position. There are two types — cutting and annealing.

The cutting type uses sliding dies made of carbide or ceramic to mechanically cut the wire. One type holds the wire captive within a hole or against an edge while an actuator slides a square-edged cutter across the top of it. Other types use two sliding cutters — some round discs, some square lozenges — in the same manner. The principle is the same for all cutting mechanisms — all components must be kept clean and the cutting edges must be very sharp. Clearance between the anvil or shearing edge must be very close or burring or flattening of the end of the wire will result. This can contribute to wire missfeeds and rethread failure.

Other threading mechanisms work by annealing and stretching. This is accomplished by applying an electrical current through a small section of wire above the upper guide. As the wire is heated to the point of melting, another mechanism pulls the heated section apart, stretching it apart so the ends are drawn to a perfect pinpoint for ease of rethreading. If you are familiar with the way a band saw blade welder works, the opposite procedure occurs in the annealing mechanism of a wire threader.

All components of *any* type of threader should be cleaned regularly, more often when using coated wires that can flake or shave and contaminate precision mechanisms, reducing threading reliability.

Air Regulator. Most auto-threaders use compressed air. Check the air pressure gauge of the regulator periodically. Check and drain the condensation reservoir weekly, or more often if large amounts of water tend to accumulate in the air lines.

MACHINE TOOL

Lubrication. This might sound obvious, but follow the recommendations of your machine's maintenance manual. With so many makes and models of equipment in place, it would be impossible to specify the types of lubricants and intervals in a generic manner. Whatever the manual recommends, make sure this is done regularly.

CONTROL/POWER SUPPLY

Power Cables. Inspect the power supply cables monthly. Check for clean, secure connections at the machine tool and at power supply terminals. Power cables should be free from breaks or kinks. *Never* attempt to lengthen or shorten them. Their length is carefully engineered and controlled because of known electrical impedance, and any modification can significantly change cutting characteristics.

Cabinet. Visually inspect all sealing surfaces of cabinet panels upon their removal to ensure an air-tight seal. There should be an "imprint" pressed into the rubber seals indicating a good seal when the mounting screws are tightened. The intake fans are powerful and should draw air from only the intake screens.

Ensure that exhaust ports are clean and free from obstructions. This will ensure adequate air circulation.

Intake Screens and Filters. Inspect and clean all intake filters and screens periodically. This interval should be determined by operators and maintenance personnel depending upon shop conditions.

Indications for more frequent attention would be if the shop environment involves other machine tools that are utilizing coolants and/or cutting fluids that can result in their vapors being drawn into these intakes. Also, grinding operations that can suspend microscopic metallic particles in the air are especially hazardous to electronic circuitry. If these conditions cannot be prevented or controlled, frequent attention to this area is highly recommended.

Circuit Boards. Most shop environments exercise reasonable control over airborne vapors and particles. In these environs, a yearly check of circuit boards is sufficient. In areas where coolant mists and/or grinding dust is prevalent, more frequent inspection and cleaning is required.

If necessary, carefully remove circuit boards *one at a time* to avoid any mixups, and gently clean any dust or carbon "fuzz" from the circuitry with a clean, soft brush.

Caution!! Never use an air hose on the boards or within the cabinet!!

Before reinstalling the cleaned boards, clean the contacts with a contact cleaner designed for this purpose. *Do not* rub them with a pencil eraser or with Scotch-brite. This can damage the fine plating and actually defeat the purpose of achieving good contact.

Disk Drives. To prevent data loss or distortion, it is recommended to clean each drive with a commercially available disk cleaner. This should be done at least once per month, and more frequently when running programs directly from disk.

Software. Make frequent backups of program disks. Aside from occasional, rare disk failure and the resulting loss of information, fluctuations of incoming power while editing or saving data can actually scramble or erase valuable programs. If this is a common occurrence (more than twice in a year), and it is not attributed to disk failure, it is strongly suggested that voltage suppression may be warranted (this is protection that transformers or "a dedicated line" cannot provide; see "Incoming Power" earlier in this chapter).

Duplicate all system disks and store them in a secure area. This will safeguard against system data loss and the resulting inability to use the machine without manually reentering data considered to be the computer's "brains" or "fingerprints."

A copy of the machine's system file or "brains" should likewise be duplicated and stored in a secure area for insurance against loss or damage to the original. An operator's copy should always be stored in or near the machine.

GENERAL

Documentation. All records, service reports, internal memos, etc., pertaining to the machine should be copied several times and stored: 1) in or near the machine itself, 2) in the Maintenance Department file, and 3) wherever the original purchase order agreement is filed.

Redundancy in this area will ensure against loss or destruction. This is to document the machine's history from the time of installation. This can provide important data that can assist in troubleshooting by indicating possible trends in machine performance, or to assist in the future trade-in or resale of the machine.

If you are lacking any book, manual, or instructions for a machine, call your distributor or machine manufacturer. If your machine is older or is no longer being made, contact someone in your area with the same equipment to see if they will give or sell you copies of their documents.

Service and Support. If at any time there is a question or problem pertaining to any aspect of your machine's operation, performance, or maintenance, do not hesitate to contact your local distributor or manufacturer. Keep telephone numbers of local service and applications engineers near the machine for easy operator access.

(*Note*: The above procedures are only guidelines, considering the numerous makes and models of equipment in the field. Therefore, these instructions have been kept very generic for obvious reasons. For specific instructions, consult the manufacturer's manual whenever possible.)

WASTE DISPOSAL

So far, the EDM industry as a whole has been lucky when dealing with waste disposal. There are a few cases where users have been fined for improper disposal of resin or filters, but, for the most part, this hasn't been a universal problem. Not yet, anyway.

If you are a shop that is producing a significant amount of metal chips from conventional machining or grinding, then you are probably already aware of some of the difficulties in disposing of spent cutting oils and coolants. In wire EDM, the dielectric water, resins, and filters can be heavily laden with waste metals and metallic ions. Some of these are worse than others, depending upon which materials you are cutting. In many states, these items cannot legally be dumped into landfills or sewers, but, by the same token, there aren't any universal guidelines for proper disposal, either. The best course of action to take is to check with the authorities in your area for specific instructions.

Right now, there is considerable activity in the filtration areas by both EDM manufacturers and after-market suppliers. This area of dielectric purification and waste disposal is reaching volatile stages as environmental concerns in the U.S. continue to increase. The days when an operator can routinely discard used filters in the dumpster will soon be over and, in many areas of the country, it already is.

Unless filterless systems can be devised, one might foresee the day when all filters will have serial numbers and registration papers just like automobiles or firearms. Suppliers and shops must either have the filter itself in their possession or a signed document stating where, when, and by whom this filter left their possession. Filters will be tracked from their initial manufacture to their final disposal or remanufacture. One can only imagine the costs and problems inherent with this type of system. We can only hope that filterless systems can be perfected, and the resulting dried "cake" from this process can be controlled and recycled much more easily.

We can't forget about resin disposal. Filter systems clean the water mechanically, by physically filtering out particulants such as metal "chips" and carbon compounds left by the wire and workpiece. Mixed-bed resins are comprised of strong acids (cation resin) and strong bases (anion resin), that are used to chemically clean the dielectric by removing metallic ions and minerals from the water. In addition to already containing these strong chemicals, after the resin's effective use, the resin beads will also contain high levels of metal and metallic compounds so its proper disposal can be a problem.

Once again, check with the local authorities in your area for the guidelines concerning proper transportation, handling, and disposal of these products. As with all waste disposal problems; don't be afraid, but be aware.

SUMMARY

As was pointed out in the beginning of this chapter, this book has not been written to frighten the would-be buyer, but rather to make the reader aware of the necessity of proper planning and preparation. There is a lot to consider, and I have tried my best to be fair and objective. While there is great potential for profit in this business (and I am quite sure all of the EDM sales personnel have done their best to assure you of *that*), in all fairness it should also be noted that there are potential pitfalls to be encountered if your planning is marginal.

If you, as a first-time buyer of a wire-cut EDM, follow these guidelines when preparing for your future EDM operations, you will find that the integration of this equipment into your operations will be much easier on your operators, your budget, your scheduling, and (indirectly) the service and application engineers of the builder of your selected machine.

While the reasons for making things easier for yourself are obvious, you may wonder how this will benefit your machine builder, and why should you care? Good question.

Answer: If you have prepared your people and facility properly, these engineers won't be unnecessarily retraining your people, performing your preventive maintenance, or chasing "ghosts" in the machine. They will now be available to help you and others with the real and sometimes unavoidable problems that you will encounter.

In closing, an EDM associate and acquaintance of mine, who runs a very successful EDM shop on the East Coast, has suggested that instead of basing EDM success upon "flushing, flushing, and flushing," perhaps it should include, "planning, planning, and planning."

I can find no argument with that.

Chapter 11

WIRE TYPES AND SELECTION

INTRODUCTION

In the early years of wire-cut EDM, choosing wire was very simple — copper wire was the only choice. Within a very short time, however, the limitations of pure or straight copper wire were discovered and various alloys were tried. Brass proved to be adequate for a long time, and today it remains the most widely used material. More recently, the increasingly difficult-to-machine alloys and the demands for faster cutting speeds brought about new wire materials that can support the high-performance power supplies available today.

Wire selection today can be just as difficult as choosing the right electrode material for vertical EDM operations. With so many different possible combinations of wire versus workpiece, there are almost as many possibilities of selecting the wrong one. Before the selection process can begin, we must first consider several basic properties of EDM wire and how they will affect our ability to wire-cut successfully.

EDM WIRE PROPERTIES

TENSILE STRENGTH

Tensile strength is the maximum load-bearing capability given to a material based upon its ability to resist stretching and breaking. It is determined by the maximum load in pounds per square inch divided by the cross-sectional area of the wire. High-tensile wire is used when using small- and fine-diameter wire to reduce wire breakage and also when cutting tall parts. High-tensile wire is also good for skim cuts, aiding in part straightness and geometric accuracy.

During wire-cut operations, the wire has a natural tendency to lag behind the wire guides. The amount of lag can vary depending on cutting speed and part thickness, ranging from a few tenths during skim cuts to several thousandths during roughing. This is called the "bicycle effect," after the way a bicycle behaves while executing a turn. I'll explain. When cutting rapidly and a turn is executed, the control arms (holding the wire guides) follow the correct programmed path but the wire itself will have a natural tendency to "cut the corner," often scrapping the part. (See Figure 11.1.) Analogous to a bicycle, the front tire is the wire guide and will follow the programmed path but the wire will behave like the back tire and will tend to take a short-cut.

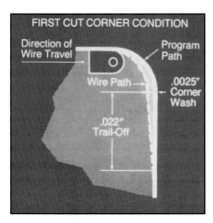

Figure 11.1   Illustration of corner washout or "bicycle effect" which occurs during high-speed machining.

Two things will aggravate this condition, fast cutting speeds and the distance between the wire guides or guide opening. Speed is perhaps the most obvious cause but taller parts also increase this tendency. To help reduce the "bicycle effect," a high-tensile wire should be used. This allows the wire to be run under higher tension, reducing the amount of "lag."

Programming strategy can also help — for instance, slowing down the cutting speed when approaching and executing a corner. Some controls have the ability to "read ahead" many lines within a program and slow down automatically, allowing the wire to catch up to the guides for an accurate corner.

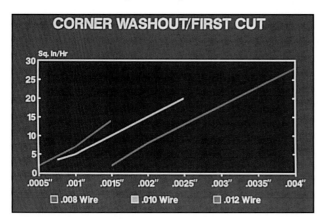

Figure 11.2   Chart showing corner washout related to speed. Larger diameter wire can cut faster so washout will be more severe. (Courtesy of MC Machinery Systems, Inc.)

Another condition users must contend with is the "barrel effect." Often referred to as "bow" or "belly," this is caused by the wire vibrating or resonating within the cut like a bowstring. This vibration is usually greatest at centerline between the guides or in the middle of the part. Where the wire is vibrating the most it is "thicker" than less-vibrating parts making the wire cut with a barrel shape, directly affecting part straightness. The taller the part is, the more likely this is to occur. Again, a higher tensile strength wire can be pulled tighter and will help defeat this condition.

Many times, tensile strength is the only criteria used in wire selection because most users believe that the higher the tension is rated, the less likely they will encounter wire breaks. Since most wire breaks are encountered well below the wire's prescribed maximum tensile strength, this obviously does not apply.

Example: Thread any brass wire on your machine and then set the correct wire tension. Let the wire run and watch it for as long as you have patience. It *didn't* break, did it? And if you don't EDM with it, it will *never* break unless you set the wire tension above the wire's rated maximum (which you wouldn't do). Obviously, high-tensile strength isn't the only factor of wire breakage.

FRACTURE RESISTANCE

Fracture resistance of an EDM wire might better be described as wire *toughness* or *resilience*. It is the ability of the wire to resist the effects of the incredibly hostile environment of the spark gap. Many factors add up to reduce the effective tensile strength of a wire. When EDMing with wire, the cross section of the wire is gradually being reduced by spark erosion as it passes through the workpiece. This alone increases the stress on the wire. At the same time, EDM craters are being burned into the wire that are usually larger than the craters on the workpiece. These craters are flaws and stress-risers that can allow cracks to propagate and cause wire breakage to occur. Too much current, poor flushing, secondary discharge, or too slow a wire speed will all further stress the wire. Eventually, all these factors will exceed the wire's toughness and it will fail. With this knowledge, selecting a wire with a high fracture resistance is your best insurance against wire breakage.

CONDUCTIVITY

This is the measure of a material's ability to carry electrical current. The higher the conductivity, the more power can be delivered to the workpiece. In most cases, this means increased cutting speed. Of existing wires, copper offers the highest conductivity, but they do have limitations due to their relatively low tensile strength. Obviously, compromises must be made in balancing the wire's flushability, fracture toughness, and conductivity. (See "Coated" Wire later in this chapter.)

VAPORIZATION POINT

In wire-cut operations, a low melting/vaporization temperature of the wire will prove the best for flushability. Remember, "flushing, flushing, and flushing?" Flushing is just as important to wire EDM as it is for sinkers, and a low vaporization point is good for several reasons.

1) The initial spark impulse vaporizes material from both the workpiece and the wire. The quicker the wire vaporizes, or *sublimates*, the faster heat will be transferred to the workpiece removing more material. The rapid erosion (within reason) of the wire is not a problem because its speed can be increased and it is continually renewed by fresh wire from the spool.

2) When the surface coating of the wire can vaporize rapidly, it protects the body, or core, of the wire by not transferring heat into it. This could contribute to an annealing of the wire that would decrease its tensile strength and cause it to stretch or break.

3) When wire *melts* instead of vaporizing, it will create chips instead of gases. These molten chips quickly resolidify in the water and, since they have mass, they take up valuable space we would rather reserve for water. When the metal is *vaporized*, gases are produced and escape from the spark gap much easier and with less influence on flushing. These gases condense in the cool dielectric but will always have a much smaller mass than a melted chip. This results in smaller chips and contaminates to lower flushing efficiency.

4) During the on-time part of the cycle, gas bubbles are constantly being created. This sheath of gas surrounds the spark from wire to workpiece. These bubbles form and grow quite rapidly only to collapse and implode during the off-time part of the cycle. When these gas bubbles can form quickly and expand rapidly, the condition displays what is called a *vapor pressure*. This rapid agitation of large volume gas bubbles growing and collapsing is an aid to flushing.

5) Because of lower melting/vaporization temperatures, the craters produced on the surface of a low temperature wire such as brass will be larger than those found on a high temperature wire such as molybdenum. This greatly improves flushing because the large craters will leave more room for the wire to act as a "linear water pump" or conveyor belt, carrying away larger amounts of contamination and damaged dielectric from the gap.

HARDNESS

The hardness or temper of a wire refers to the wire's *ductility*, its ability to undergo elongation — not its tensile strength. A wire's hardness or softness determines how much elongation of the wire can occur. In this category, EDM wires are called either soft or hard.

1) Soft wire is malleable and can undergo elongation which allows it to stretch and deflect without breaking. Elongation of soft wires averages around 15 to 17%, and some even exceed 30%. Soft wire should be used when taper-cut-

ting beyond a 7° angle because it will elongate as it passes across the guides instead of taking a "set" or displaying memory like a hard wire does. ("Set" as it is used in this case is the same term used in die-making. You can "set" a part as in *coming*.)

2) Hard wires display properties that could be called "memory." If a hard wire is bent, stretched, rolled around a small diameter spool or drawn across an edge, it cannot readily elongate so it will tend to remain in that shape. Most hard wires cannot stretch more than 2 or 3%, making it a "stiffer," straighter wire. That is why hard wires are better suited for use with automatic wire threaders. When a hard wire leaves the upper guide, it "remembers" that it is straight and it will remain so within the jet of water, assuring higher threading reliability.

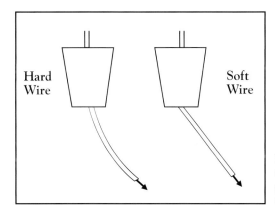

Diagram I.   An exaggerated profile of a hard and soft wire entering and exiting a typical guide under high tapering conditions. (Courtesy of Gisco Equipment, Inc.)

Explanation: Except in machines that have gimbaled or articulated heads, during taper cutting operations, the wire is forced across the radius of the guide. This compresses, compacts, and burnishes the inside of the curve while stretching and rarefying the structure of the outside curve. The hard wire will work-harden on the inside curve of the wire that now has densely packed molecules. In this situation, a hard wire will take a "set" and will want to continue curling. A soft wire is much more malleable and has the ability to elongate easier, allowing it to stretch and "flow" across the guide radius and undergo taper-cutting without "setting" or curling.

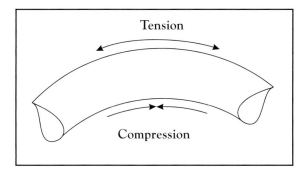

Diagram II.   Tension/compression loads caused by bending the EDM wire. (Courtesy of Gisco Equipment, Inc.)

EDM WIRE TYPES

As stated earlier, wire selection was easier in the early days of EDM — there simply were not a lot of choices. Today is a much different story; now we must select from different wires that have specific properties keyed for use against different workpiece materials. We also must contend with the knowledge that some EDM generators perform better with a particular wire brand or alloy. Let's start at the beginning.

Figure 11.3   There are many different sizes and kinds of wire available. (Courtesy of Gisco Equipment, Inc.)

COPPER WIRE

When wire-cut EDM was in its infancy, copper wire was a natural choice. Pure copper wire for EDM has a very low tensile strength (34,000 to 60,000 psi), and is subject to elongation and excessive breakage. Also, if the most important aspect of wire-cut EDM is flushing, then perhaps the greatest shortcoming of copper wire is that it has poor "flushability" due to its high vaporization temperature when compared to brass wires (brass wires contain zinc which lowers the vaporization temperature).

Although copper wire has excellent electrical conductivity, 100% IACS (International Annealed Copper Standard), it also has high thermal conductivity. This inhibits cutting speed because it sinks heat away from the cut and into the body of the wire, which in turn slows the melting process and reduces efficiency because larger amounts of heat are transferred to the wire instead of the workpiece.

BRASS WIRE

Plain brass wire is by far the most widely used wire type. Brass is made by alloying copper with zinc. The most common type is "yellow brass" that has a copper/zinc ratio in the ranges of 65/35 to 63/37 and is available in tensile strengths ranging from 50,000 to 145,000 psi. Compared to copper, brass wire provides higher tensile strengths and a lower vaporization temperature. This allows smaller chips to be formed and also allows larger craters to be formed on the surface of the wire to carry them away.

High-tensile brass wires are made by alloying brass with small amounts of aluminum or titanium, giving it a makeup of 65/33/2 Cu/Zn/Al. Tensile strengths can range from 140,000 to 173,000 psi, but flushability will suffer as strength increases.

SPECIALTY WIRE AND "COATED" WIRE

In this context, "coated" has been set in quotation marks because it is being used to group together and describe several different wire types. Wires within this group can be called "plated," "coated," "stratified," and "diffusion annealed." The terms coated and plated describe the same type of wire, while stratified is another term commonly used for diffusion annealed wires. All are correct, if not confusing.

All coated or plated wires have a thin coating of pure zinc that is applied by an electroplating process using special equipment to ensure a homogenous coating of zinc with a uniform thickness that is reasonably concentric to the core diameter. Diffusion annealed wires have a thicker coating of zinc applied to the wire's core and it undergoes a heat treating process that changes it to brass and fuses this outer coating to the core.

These wires came into being with the logic of: "If I need the tensile strength of one material, and the conductivity or flushability of another, why not coat or plate one with the other?" That is exactly what wire manufacturers have done.

In this chapter, I am generalizing for discussion's sake when I refer to "zinc" wire. Alloy compositions used in "plated" and "stratified" wires can vary widely, but each of them involves pure zinc or higher percentages of zinc, and, therefore, my generalization.

Zinc has become a widely used wire coating and alloy material because it provides faster cutting speeds and reduced wire breakage. This is because zinc has a much lower melting/vaporization temperature than brass. In fact, the zinc coating will boil at 906°C (1663°F), long before brass can melt at 930°C (1706°F). (See Figure 11.4.). This means that the lower melting temperature of zinc will allow

MATERIAL	MELTING TEMP. °C / °F	BOILING TEMP. °C / °F
Aluminum	660° / 1220°	2450° / 4442°
Brass	930° / 1706°	
Copper	1082° / 1980°	2595° / 4703°
Graphite	NA	3500° / 6332°
Magnesium	650° / 1202°	1107° / 2024°
Molybdenum	2625° / 4757°	5560° / 10040°
Titanium	1820° / 3308°	3260° / 5900°
Zinc	419° / 786°	906° / 1663°

Figure 11.4   Temperature chart for wire components.

larger craters to be formed on the wire, improving flushing. It also helps ensure that as long as there is a layer of zinc to "boil away," the brass core can never get hotter than the vaporization temperature of zinc and burn through. Wire breakage is reduced and cutting speeds go up.

We can understand how improved flushing is a major reason zinc wires cut faster, but why do they break less often? I would call this "core protection." The coated or stratified structure protects the inner core of the wire similar to the way a double-boiler cooking pot prevents burning heat sensitive food. A double-boiler is a set of two pots with one nested inside the other. The larger, outer pot sits on the stovetop and contains boiling water. The bottom of the smaller pot is immersed in the boiling water and heat is transferred to the food in this manner.

Why do this? Since boiling water cannot exceed 100°C (212°F), it cannot burn the heat sensitive food. The analogy to EDM is that no matter how hot the flame (EDM spark), the water (zinc coating) in the bottom pot (the wire's surface) can never exceed its boiling point so it cannot burn the heat sensitive food (the wire's core).

In addition to improving flushability through quicker vaporization and high vapor pressure, the zinc coating, in effect, also "protects" the brass core. This is an important principle of coated-wire behavior, providing that the wire speed is fast enough and the coating is thick enough to not be wholly eroded away before it exits the workpiece. This applies to any coated or stratified wire with a lower temperature external coating. There are many types of these wires and we will examine them next.

COATED COPPER CORE WIRE

Older generators (and certain Swiss made generators) favor this zinc coated wire because of the high conductivity of its copper core. Although this wire gives much better cutting performance than pure copper, the copper core can limit its accuracy because its low tensile strength prevents it from being "pulled tight" for straightness and accurate skim cuts.

For faster cutting, a similar wire was developed by adding a small amount of magnesium to the copper core. It is then plated with zinc and heat treated within an oxygen atmosphere which causes the hot outer surface of the wire to oxidize, and the wire is now covered with a very thin layer of zinc oxide. This is done for several reasons.

1) The zinc oxide surface is "more open" when compared to a drawn surface. It has very small micro-cracks that help carry contamination from the gap area, improving flushing.

2) The oxidized layer is harder than zinc. This helps the wire slide across the carbide contacts and wire guides with less friction. It also prevents flaking which can contaminate wire guides and precision threading mechanisms.

3) The zinc oxide wire surface is only semi-conductive. Although its not supposed to, the wire *does* brush and contact the workpiece because of mechani-

cal vibrations, flushing pressure and the sheer dynamics of the spark itself. As any operator knows, contact between the wire and workpiece results in a short, and the servos will retract the wire from the workpiece until the short condition is eliminated. These high-speed, intermittent contacts aren't severe enough to make the servos back up but they are enough to collectively slow the cut down. These brief contacts could be called "mini-shorts." Because the wire's semi-conductive oxide layer helps "mask" most of these mini-shorts, this essentially "desensitizes" the servo system, allowing for more aggressive cutting than could probably occur otherwise.

This wire is good for speed cuts since wire tension is typically less during roughing, giving it more resilience and resistance to breakage. Flushing is also improved.

Another coated copper wire is made with a brass coating of 50/50 Copper/Zinc (hereafter, Cu and Zn) that, after heat treating, melds with the core and leaves the wire surface with an open or porous outer structure that aids in flushing. The wire then undergoes a final drawing that further compacts and presses the diffused coating into the core. Diffusion annealed wires are good for taller parts because the outer coatings will be thicker, almost 35μm per side. This helps the outer surface to last longer on tall parts.

COATED BRASS CORE WIRE

These wires typically have a brass core within the range of 63% Cu and 37% Zn with a thin coating (2 to 3 μm) of pure Zn that is deposited by plating or galvanizing. This type of wire is good against most materials including carbide. However, the extremely thin coating limits its use against tall workpieces because the zinc coating will be consumed before the wire exits the workpiece.

There is a brass cored wire commonly referred to as "brass on brass." It has a diffusion annealed brass coating of 50/50 Cu/Zn and a core of 80/20 Cu/Zn. Note the higher percentage of Zn on the outer surface that lowers the vaporization temperature and aids flushability.

There is even silver coated brass wire. Silver offers higher conductivity over even copper: the IACS rates copper at 100% and silver at 105%. This wire provides faster cutting speeds because of the efficient transfer of energy from the carbide contacts. It costs less than one might expect (the coating is *very* thin), and is priced somewhere between plain brass and coated wires.

MOLYBDENUM WIRE

Often called, "moly" (pronounced "*molly*") wire. This is a very high strength wire, with tensile strengths in excess of 275,000 psi, over 100,000 psi more than any plain-brass wire. While it offers a very high tensile strength and high resistance to breakage, moly wire has a very high melting temperature of 2625°C (4757°F). Its high boiling point of 5560°C (10040°F) causes it to cut very slowly because of its poor flushability and low vapor pressure. The EDM craters formed on moly wires are quite small and can't contribute much to flushing.

These wires are generally used in the small, (0.006 to 0.004"), and fine, (<0.004"), diameters to meet blueprint requirements of near-sharp, inside corner radii. Because of its high tensile strength, moly wire aids in maintaining wall straightness and reduces the amount of breakage found with small and fine wires.

Note: Use of this wire must be application-specific because of its high cost and slow machining times. Use of moly wire has declined except in the finest wire sizes because improvements in high-tensile brass and high tensile coated wires have made them very successful in small-wire applications down to 0.004", and at a much reduced cost over moly.

COATED MOLYBDENUM WIRE

We know that moly wire is incredibly strong, but we also know that it cuts slowly. About ten years ago, an interesting twist in wire development borrowed existing technology from vertical EDM machines by applying a thin layer of graphite to a moly wire. Graphite, being a metalloid, does not melt but it sublimates — it changes directly from a solid to a gas. With little or no transition time in changing from a solid to a gas, it delivers a much more efficient spark to the workpiece. Further, in a water dielectric, the graphite coating will oxidize and produce gases instead of "chips." This improves flushing conditions because there is much less debris remaining to contaminate the gap.

COATED STEEL CORE WIRE

There are special wires available that have a steel core (for high strength) coated with Cu (for conductivity), and then coated with brass (for cutting qualities). Two (of three) wire labels show that the coating thickness, tensile strength, and conductivity rating of these wires are similar to those of brass diffusion annealed wires, but steel cored wires have the advantage of much better resistance to breakage because of their high fracture resistance.

Tensile strengths for these wires are determined by the amount of carbon in the steel core. Steel core wire with a high carbon content will produce a stiffer, more "springy" wire. Steel core wire would be a good choice when attempting extremely tall parts, parts requiring good straightness, and where poor flushing conditions exist.

WIRE QUALITIES

The wire must meet the following standards to ensure success:

1) **Material Quality.** Wire must be made of the highest purity alloy(s) available for this purpose. The drawing and heat treating processes must produce wire surfaces that are smooth (in the case of plain brass wires) and uniform (for "coated" wires). The wire surface must be free from any flaws that can propagate cracks under tension and cause wire breakage. In addition to start-

ing with high quality raw materials and paying strict attention to quality control during the many drawing and annealing steps of manufacture, the *cleanliness* of the finished product is of the utmost importance. A dirty wire will contaminate rollers, autothreaders, and wire guides, delivering poor performance and increased maintenance.

2) **Accuracy**. The wire must have a precision uniform diameter. Besides the obvious effects on accuracy, this can affect the performance of the machine's wire-delivery system and automatic wire threader. The standards for EDM wire's physical dimensions are quite high, with the standard tolerances for plain brass wire being ±0.000040", and ±0.000060" for plated wires.

3) **Tensile Strength**. The tensile strength must match the application or inaccuracies and excessive wire breakage can occur. Wire breakage is typically not directly related to tensile strength and this is seldom a problem because all quality wires have a generous safety margin in their ratings.

4) **Hardness**. The hardness or ductility of the wire must be uniform and consistent. While tensile strength is important, if the wire is too hard or brittle, it cannot withstand shock and will break. Different hardnesses are used for different applications. For example: when cutting more severe tapers (7° +), using a soft wire is advised because pressure across the guides will stress one side of the wire more than the other and result in a "curling" effect. Machines using automatic wire threaders will prefer a hard wire with no "memory" for the most reliable threading.

5) **Spooling**. The winding of the wire, or spooling, must be precise and uniform. Producers go to great lengths to ensure the spools are wound without snags, overlapping, or pinching winds or variations in tension. Observation: The best wire in the world won't help a bit if it doesn't unwind from the spool.

6) **Plating or Coating** (where applicable). Any plating or coating on a wire must be uniform in depth and density. If the wire is improperly plated, it can have thin spots resulting in erratic cutting performance, unnecessary breakage, and potentially inconsistent dimensional and surface finishes. If the plating or coating of the wire has not been processed correctly, the zinc surface can flake or scale off as it passes through the rollers and guides, causing additional wire breakage and wire threading problems.

7) **Packaging**. While not exactly a quality of the wire itself, packaging is important in assuring that all the above characteristics are used to their best advantage. All spools should be sealed in plastic bags to avoid oxidation and contamination, and they should be unwrapped only when ready to use. Many wires are shipped "nested" in their own Styrofoam boxes for protection. When purchased in bulk, wire stock should be dated and rotated to assure quality.

WIRE TERMINOLOGY

Before we proceed any further, there might be something for you to keep in mind; terminology of EDM wires may vary from manufacturer to manufacturer. This means that one maker can consider a certain grade of wire to be "hard," while another maker might consider that same grade to be "half-hard." Therefore, their recommendations to you can unintentionally confuse or conflict. I'd suggest obtaining or making a wire matrix or data base showing all wires and their compositions for a better comparison. Terminology may confuse you but numbers shouldn't.

TROUBLESHOOTING

Now that we have all these different types of wire, what do we do with them? What do I use when I want to cut fast? What about tapering? What do I do when I keep experiencing excessive wire breaks? First, let's make sure we know the correct limits of wire tension. The chart in Figure 11.5 shows the maximum wire tension, in grams, that different wire diameters can physically withstand. *These are NOT machine settings.* These figures are for reference only as they are direct conversions and could not realistically be used while machining.

The chart in Figure 11.6 gives some general guidelines for wire selection and troubleshooting.

GUIDELINES

Increasing the fracture toughness of the wire is perhaps the single largest "fix" for reducing wire breaks. For increasing speed, use coated wire, although zinc plated wires will show their limitations in thicker workpieces because the zinc coating is very thin and will be consumed before it exits the workpiece. For thicker workpieces, go to a diffusion annealed wire that will have a thicker coating of brass. This ensures that the coating survives all the way through the cut while improving flushability. For taper cutting, use softer wire to prevent the wire from curling as it passes through the offset guides. Automatic wire threaders prefer harder wires to aid in threading through small start holes. For high accuracy applications, increas-

Wire Dia.	60,000	70,000	PSI 90,000	105,000	130,000	140,000
0.004"	270	300	NA	450	650	750
0.006"	600	650	NA	1000	1500	1750
0.008"	1100	1200	1650	NA	2650	NA
0.010"	1700	1850	2550	NA	4200	NA
0.012"	2400	2700	3700	NA	6000	NA

This conversion chart is for reference only. These values are correct in their conversion but are too high to be practical for cutting.

Figure 11.5   Maximum wire tension in grams.

ing the wire tension will be a big help, using high tensile brass, steel core, or molybdenum wire.

One more item can be listed here concerning wire breakage that wire selection cannot always solve. This concerns wire breakage on older machines. Some older machines have antiquated servo systems that may not be able to provide the servo response necessary to prevent repeated wire breakage on a given job. Many times a speed reduction will be the only remedy.

PROBLEM	REMEDY*	WIRE
To Reduce Wire Breakage.	Use larger diameter wire. Decrease roughing tension. Use "soft" servos. Increase tensile strength. Increase fracture toughness.	Al or Ti/brass alloy. Coated steel core.
To Increase Cutting Speed.	Use larger diameter wire. More power. Better flushing.	Diffusion annealed copper. Zinc coated brass.
To Cut Steep Tapers.	Increase ductility.	Soft brass. Zinc coated brass. Diffusion annealed brass.
To Cut Thicker Workpieces.	Use larger diameter wire. More power. Better flushing. Increase tensile strength.	Diffusion annealed copper. Diffusion annealed brass. Coated steel core.
To Increase Straightness.	Increase tensile strength.	Al or Ti/brass alloy. Moly wire. Coated steel core.
To Increase Accuracy.	Increase tensile strength. Increase fracture toughness.	Al or Ti/brass alloy. Coated steel core.

Actual adjustments of current, servo speeds, flushing pressure, etc., are not noted here but may be necessary to realize the full potential of the troubleshooting process.

Figure 11.6   Guidelines for wire troubleshooting.

SUMMARY

Selecting the correct wire for the job comes from a basic knowledge (which we have tried to establish here), experience, and the recommendations of the wire manufacturers, distributors and the builders of the machines themselves. Experience alone used to suffice, but in the past few years, wire technology has improved so much and so fast that we must rely, in part, on the advice and suggestions of the suppliers. This is especially true with the newer wire-cut machines as their power supplies, flushing capabilities, tapering and tensioning devices, and threading mechanisms can be significantly different, especially between machines of Swiss and Japanese origin.

Another excellent source of valuable information can be the applications engineers employed by the builders of your equipment. Many times, they will be sent samples of new types of wire for "beta" testing before its release to the public. In this manner, applications engineers can learn each wire's cutting characteristics and develop efficient conditions and settings for that particular power supply. These same engineers should also be able to recommend wires and cutting conditions for the older machines that are still in the field.

There are also many EDM seminars and technical conferences held throughout the country, with speakers and specialists who possess information or literature that might help you. Aside from their actual presentations, most of these individuals welcome the opportunity to speak directly to others during breaks in these seminars, or after hours, and would be glad to discuss any questions or particular problem you may be having.

One final item that is often overlooked is maintenance. If your equipment is in disrepair or operating at less than optimum conditions, then *any* objective evaluation of a particular wire's performance will be difficult, if not impossible.

Wire machines require regular preventive maintenance (see Chapter 16), without which performance will drop off rapidly no matter *what* kind of wire you are using. Cleanliness is mandatory. Check the power feed contacts, guides, bushings, rollers, and power cables, daily. Don't forget to monitor the cleanliness and condition of the water, and check the condition of the filters and resin regularly. With wire machines running unattended longer and longer these days, we can easily take for granted just how many hours they actually *are* working. Charting these hours and scheduling regular preventive maintenance will go a long way in assuring optimum, or at least *predictable*, cutting performance, not to mention prolong the life of other consumables and the machine tool itself. Do this. Then and only then can a fair evaluation of any wire's performance be made.

Chapter 12

WIRE FLUSHING

INTRODUCTION

Everyone has already heard, perhaps more often than they would like, that the key to successful EDMing is "flushing, flushing, and flushing." Even if you are tired of hearing it, this statement is very true. The best efforts of the best operator, on the best machine, using the best dielectric oil and electrode material can be rendered ineffectual and inefficient without adequate provisions for flushing.

But wait. Electrode material? Dielectric oil? If we are talking about electrode materials and EDM oil, we are obviously talking about sinkers, or vertical EDM, right? Right.

Question: But what about wire EDM? Isn't flushing just as important for wire-cutting? And if it is, then why don't we hear or read more about it?

Answer: Yes, *without a doubt*, flushing is just as important to wire cut operations as it is to vertical EDM. As for the second question, "…why don't we hear or read more about it?" Sorry, I don't have that answer.

What we will do is examine flushing and its importance from the wire-cut point of view. For review, or for those who missed it, Chapter 4 on flushing covered the theory and practice of flushing, but was specific to vertical EDM operations.

The theory of flushing is the same for sinker or wire, but there are obvious differences in implementation. We will examine the importance of flushing to wire-cut operations and, like the sinker chapter, offer some tips and tricks.

THEORY

To review, flushing is the process of introducing clean dielectric fluid into and through the spark gap. This serves several purposes. Flushing:

1) introduces "fresh" dielectric to the cut for reionization,

2) flushes away the "chips" and debris from the spark gap, and

3) cools the electrode (or wire) and workpiece.

As described in the chapter covering EDM theory, we must provide enough off-time to allow the dielectric to reionize or "recover." Rather than rely entirely upon the dielectric strength of the oil or water, we can reduce the recovery time by diluting or replacing the damaged dielectric in the spark gap by forcing clean, filtered fluid through it. The quicker the exchange of "fresh" fluid for "tired" fluid, the shorter the duration of off-time. This will increase the speed and efficiency of the cut, while reducing wire breakage.

FLUSHING AND CUTTING SPEED

Over the past 15 to 20 years, wire-cutting speeds in tool steels have increased from just 1 sq. in./hr. to over 30 sq. in./hr. Titanium will machine even faster and, in aluminum, speeds of over 80 sq. in./hr. have been achieved. Much of the credit for this improvement in machining speed must go to the engineers and theorists who are constantly testing and improving the high-tech circuitry found in today's modern generators and power supplies. Also, great strides have been made through the years by wire manufacturers, improving the cutting properties and flushability of the wire electrode itself, and this has also contributed to the increase in cutting speed.

Despite the advances made in machine and wire technology, the increase in cutting speeds would be substantially lessened without the concurrent improvements in flushing. Improvements in flush cup design and increased flushing pressures are largely responsible for *delivering* the improvements made in generators and wire. Much like a race car, no matter how much power you have, if it can't be delivered or transferred to the ground (presented to the workpiece), it won't do much good. High pressure flushing is part of the solution.

Most newer machines provide flushing pressures up to 300 psi. This provides the pressures required to force chips and contaminants from the gap while introducing "fresh," clean, and cool dielectric for continued high-speed machining.

FLUSH CUPS AND NOZZLES

A flush cup or nozzle is a plastic or ceramic device that is positioned over the wire guide and through which water is forced to flush chips and debris from the spark gap (see cut-away view, Figure 12.1).

The closer the opening and sealing surfaces of the nozzles are to the workpiece, the better the flushing will be because more water will be forced to enter the narrow spark gap and kerf left by the wire. As long as your workpiece surface is smooth and parallel, strive to maintain a workpiece-to-nozzle distance of 0.002" to 0.005" for the best forced flushing (see side view, Figure 12.1). Whenever we have the good fortune of cutting a plate of material that is smooth, flat, and parallel, we

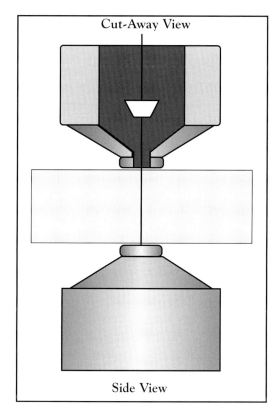

Cut-Away View

Side View

Figure 12.1 A flush cup
(with top cut-away) viewed in profile.

have a much better potential to wring the most efficiency from any given machine.

To maintain the greatest flushing efficiency possible, the physical condition of the nozzles is important. Inspect them often for chipped edges of the sealing surfaces and for cracks in the body. Chipped edges can occur from nozzle collisions with the workpiece, tooling, or clamps, or from slugs and slivers being allowed to fall free, striking the lower nozzle. Although these nicks and chipped edges may seem small, they can allow water to deflect and escape from the sides, away from the desired direction of flow, lowering the pressure in the cut.

Sometimes small cracks are difficult to find because they leak only during high-pressure flushing conditions when the deflective flushing "hula skirts" or splash diffusers are in place. Remove the suspect flush cup and tap it sharply with your fingernail or a pencil. A cracked or damaged nozzle will have an obvious sound and "feel" to it — it will not "ring." Keep spare nozzles on hand, and at the first indication of damage, replace them. Don't make the mistake of subscribing to false economy. Cracked or damaged flush cups can reduce cutting speeds as much as 50%. The minor cost of a replacement flush cup is easily justified by the major benefits of maximum machining speed and reduced wire breakage.

Optimum flushing conditions occur when:

1) the workpiece is flat, smooth, and parallel;

2) the flush cups are in good condition;
3) the flush cups are positioned as close to the workpiece as possible; and
4) the program path does not cut near the edge of the part.

But what happens if our workpiece is shaped with steps or shoulders on it that prevents us from setting the nozzles a few thousandths away from the workpiece? In this case, the part can still be machined by setting the guides and nozzles as close as possible to the maximum workpiece thickness. Through the thinner sections, we will suffer reduced cutting speed and increased wire breakage due to poor flushing, but in most cases, the job can proceed. Sometimes, due to part shape or configuration, our flushing provisions can fall short. This situation brings us to the next step toward improving flushing conditions.

AFTERMARKET FLUSH CUPS AND NOZZLES

HIGH-PERFORMANCE FLUSH CUPS AND NOZZLES

One of the least expensive yet most valuable enhancements to wire-cut operations is the "high-performance" flush cups and nozzles that are available from several aftermarket suppliers, or that can be custom made according to your needs. Some are for increased speed, others are for improved edge starts, while others provide a mini "traveling subtank" allowing a die opening, for example, to fill with water for improved flushing while skimming. A single device of this type, in the right application, can double and sometimes triple cutting speeds or improve workpiece finish and accuracy.

Factory or OEM (Original Equipment Manufacturer) nozzles work fine most of the time, but again, like automobiles, there are occasions when "stock" (meaning "stock performance") isn't good enough. We would like to "tweak" things a bit to maximize a performance gain. Many times, an inexpensive modification to an existing nozzle can make the difference between profit and loss or success and failure. Let's examine a few of these modifications.

Most manufacturers can provide nozzles with small or large openings depending upon their intended use. Large openings provide high-volume, low-pressure flushing typical of edge starts and/or skim cuts. Small nozzle openings provide high-pressure flushing for a concentrated flow to efficiently remove contaminants from the gap during high-speed, full-width, roughing cuts.

If the builder of your machine does not provide flush cups with large and small openings, they can easily be made. Larger openings can be made by drilling out smaller ones. Smaller openings or specials can be made by plugging a large opening with a nylon or delrin-type material and redrilling a smaller opening. (When plugging a nozzle for redrilling, it is a good idea to counterbore the hole from the backside and make a shoulder on the plug to prevent it from blowing out under high pressure.)

Sometimes we have difficulty cutting parts that have premachined or cast recesses or counterbores in them. This causes problems because the nozzles can only be set at the thickest plane to prevent crashing into the sides of the recessed detail. This causes the flushing pressure to be diffused or deflected before it reaches the cutting area.

In this case, "extended" nozzles can be made or bought to provide more precise and efficient flushing in these recesses and counterbores. Without them, flushing would be poor at best. With them, although the wire guides cannot be positioned any closer to the cut, the extended nozzles will deliver a much more concentrated flow of water directly to the cut, improving cutting speed and reducing wire breakage (see Figure 12.2).

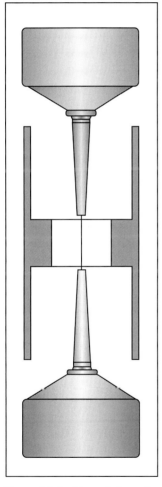

Figure 12.2 Extended nozzles can be used for more precise flushing.

When cutting steep tapers with standard nozzles, the wire is drawn at an angle, but the water will continue to flow vertically from each nozzle opening, wetting only the top and bottom of the cut. This will result in decreased cutting speeds, irregular finishes, and an increase in wire breakage.

Taking this extended nozzle concept to the next level is an articulated or "wiggler" style nozzle. This type of nozzle uses a ball and socket design that will allow the nozzles to be "aimed" parallel with the wire as in Figure 12.3. This directs the flow of water directly onto the cut, improving flushing and machining performance.

Another very versatile type of flush cup offers the ability to have a "traveling subtank" for die openings up to 4 inches. This thin, saucer-like disk can be made or purchased, and when set within a few thousandths of the part bottom, will effectively seal the die opening. (See Figure 12.4.)

Using a large-diameter upper nozzle and low-pressure water flow, the die opening will fill with water and allow submerged skim cutting with very little water pressure to reduce unnecessary hydrodynamic influence that can affect finish and accuracy. When using an automatic wire threader, this "traveling subtank" arrangement provides unattended submerged skim cutting as it moves from detail to detail. But what if our part has steps or shoulders on it that prevents us from setting the nozzles a few thousandths away from the workpiece? In this case, the part can still be machined by setting the guides

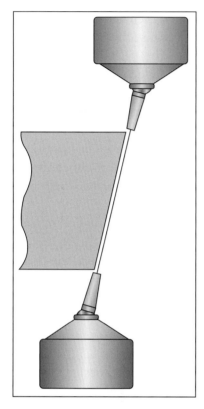

Figure 12.3   Articulated nozzles incorporate a ball and socket design for flexibility.

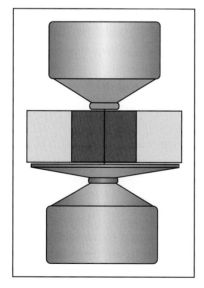

Figure 12.4   A "traveling" subtank fits below the part and effectively seals the die opening.

and nozzles as closely as possible to the maximum workpiece thickness. Through the thinner sections we will suffer reduced cutting speed and increased wire breakage due to poor flushing but, in most cases, the job can proceed. In certain shapes and configurations, all of our tricks and remedies to improve flushing can fall short. This brings us to the next step for improved flushing.

SUBMERGED MACHINING

When faced with wire-cutting stepped or irregularly shaped parts, rounds, tubing, and interrupted cuts, submerging the part in the dielectric can greatly improve flushing conditions. On round shapes, the water pressure will be deflected by the diameter of the part. It is obvious that due to the round shape, positive "sealed flushing" is impossible.

Square and round tubing will further compound flushing difficulty. (See Figure 12.5.) On multilayered or interrupted cuts such as encountered when wire-cutting the reader-head for a disk drive (see Figure 12.6), the inner levels of material will "starve" or cut "dry" in comparison to the outer layers.

Solution? Submerge the entire part in dielectric. For pure speed, nothing is better than sealed, high-pressure flushing, but, we don't always have this luxury. Usually, when cutting submerged, flushing pressures will seldom approach the pressures found in speed cutting, but at least we are assured the entire cut remains wet, providing reduced wire breakage and a substantial increase in your profit margin.

WET OR DRY

To submerge or not to submerge, that is the question. Almost all wire EDM builders provide wire EDM machines with submerged cutting capabilities. Some people insist that sub-

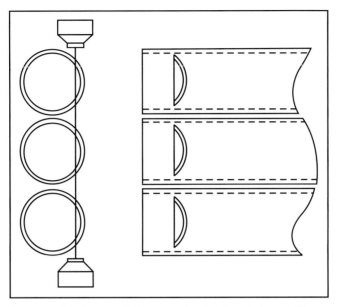

Figure 12.5   On round shapes, sealed flushing is impossible.

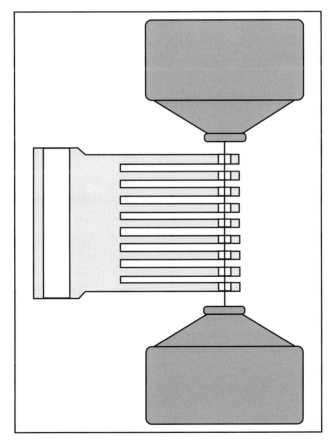

Figure 12.6   On multilayered cuts, the inner layers will "starve" or cut dry.

merged cutting is the single greatest answer to all wire EDM problems, and recommend a submerged machine in every case. Others dismiss submerged cutting as being only a marginal solution. In actuality, it all depends on the application.

Depending on its geographical location, the average wire job shop will encounter between 10 and 20% of their general workload that would require submerging. This criterion can be an important part of a shop's machine selection. One must consider the type of work they wish to process or pursue.

Question: If only 10 to 20% of my work requires submerging, should I purchase a machine that is dedicated, or 100% submerged?

Granted, all submerged machines can be used either "wet" or "dry." The limitations of a dedicated, submerged machine could appear when a job or workpiece is larger than the rigid worktank would allow. Maybe not often, but it *is* something to think about. Dedicated production or long-running jobs that require submerging, in this case, would simplify machine selection. Otherwise, machine selection for average job-shop-type operations must be a case-by-case evaluation, based on past work experience or the type of work that is to be pursued in the future.

All builders today offer submerged wire-cut machines, but what if a job comes along that should be submerged but your machine is nonsubmerged, or if you have an occasional submerged job that, by itself, would not justify the purchase of a submerged machine? In these situations, one can purchase or make a "portable" submerged worktank. These devices can provide the occasional submerged cutting capability without the large investment in a dedicated, submerged machine. Also, a removable subtank will not limit the machine to parts small enough to fit within the physical constraints of a rigid worktank.

AFTERMARKET SUBMERGED WORKTANKS

There are several styles available, and these devices offer the unique ability to quickly change a "dry" machine into a "wet" one in a matter of 15 to 20 minutes. These units are typically placed on top of the machine's existing worktable or pedestals. Some utilize a "sacrificial plate" of sheet aluminum as the bottom of the worktank. They are called "sacrificial" because when the first cut is made, the plate is also cut. Except in long-running production jobs where the wire path never changes, this plate will eventually have to be replaced. Another advantage that this alternative offers, besides submerged cutting when needed, is that the lower nozzle is below the sacrificial plate, so slugs can be dropped without concern for wire breakage, jammed nozzles, or interference with the lower arm.

Another similar type of portable, submerged tank is more complex, having a "traveling bottom" or "sliding window" that is made of sliding components that allow the lower guide and flush cup to extend through the bottom of the tank and move the nozzle opening as the machine travels in X- and Y-axes.

Smaller portable units can consist of just the portable tank itself, while larger systems must be augmented with auxiliary reservoirs, pumps, and valving to sup-

Figure12.7 Example of a submerged cutting within a fabricated "flask" or "jacket." This machine has been specially modified to cut three parts simultaneously. (Courtesy of Spectrum Manufacturing, Inc.).

port the additional water requirements. Should you consider one of these units, make sure that its builder is aware of all the details and requirements of your application. Other submersible tanks can be "made-to-order" units. These are usually for production applications and are typically "flasks" or "jackets" to form fit the part or multiple parts, submerging it/them in a minimal amount of water (see Figure 12.7). Not only can this eliminate the need for extra pumps and plumbing, but it will also reduce the amount of water-weight that must be moved around the table.

"SPECIALS"

CASE STUDY

Some jobs require a "fix" or "fixes" that are beyond the scope of special flush cups or simple submerging. A good example of this would be the problems faced by a premier EDM job shop in Southern California, when they took on an extremely large center-hub (rotor) for a new type of jet engine made from a titanium forging with a major diameter of over 34-inches and a weight of 790 pounds.

The traditionally familiar "fir-tree" root of turbine blades are not used on this new style engine. Instead, they are configured in such a way that wire must pass through several varying sections of thickness ranging from

Figure 12.8 View of titanium aircraft engine rotor after wire cutting is completed. (Courtesy of Swiss Wire EDM, Inc.).

1" to 13". Roughing this part was not extremely difficult because with a programmable Z-axis, good, sealed, high-pressure flushing was possible throughout most of the cut. Skim cuts were much more difficult because of the height of the part, complicated by strict print callouts of straightness, contour, and dimensional tolerances with finish requirements requiring only 0.0003" of recast.

Figure 12.9   A total of twenty-two slugs are removed from the forging, each weighing almost thirteen pounds. (Courtesy of Swiss Wire EDM, Inc.)

Figure 12.10   One of the slugs is used to fabricate a flushing fixture to allow submerged skim cutting. Assembled view flushing dam. (Courtesy of Swiss Wire EDM, Inc.)

To supply this part with such demanding criteria, yet also within a reasonable time frame, a flushing fixture, best described as a "flushing dam," was fabricated from wire-cut materials. For the "body" of the flushing fixture, the actual slug from the very first cut was used. To this, aluminum plates were bolted to both top and bottom to ensure maximum sealing capability, allowing water loss only through the precut wire-cut slot. Auxiliary flushing hoses were attached to fittings on the back of the fixture to supply the additional volume of dielectric required to flush the spark gap and keep the fixture full.

First, the entire part was roughed out. This allowed the part to move around as stresses were relieved, permitting the part to "relax" prior to finishing. Then the "stuffer plate" assembly was dropped in and secured. Next, skimming was carried out until the detail was finished and within all tolerances. It was then moved to the next detail in turn until the entire part was finished. Specifications of this cut are as follows:

1) Roughing: Wire = 0.012" diameter Bronco X, removing a total of twenty-two 12.5 lb. slugs at daytime speeds in excess of 27 sq. in./hr. Nighttime and unattended speeds were reduced 20% to ensure zero wire breakage.
2) Finishing: Wire = 0.012" diameter brass, using special flushing fixture, skims to a straightness tolerance of ± 0.0003" straightness from top to bottom (over 13.5"!) and ± 0.0006" overall part contour. Recast must be held to 0.0003" maximum.

Figure 12.11   View of flushing dam installed on part. Water fills narrow space between the slug and the part, allowing much faster, straighter skims with no wire breaks. (Courtesy of Swiss Wire EDM, Inc.)

Careful planning and provisions for flushing allowed the shop to wire-cut this part 33% *faster* than previous methods in spite of the part size and the daunting tolerance requirements. Their final results were improved: speed, finish, accuracy, and unattended machining. Almost all were the results of improved flushing. Let's examine why.

Improved flushing enhances EDM machining for the following reasons.

1) It is uniform throughout the cut because it is submerged. On such a tall part there is no danger of the part "cutting dry" and burning through the wire. Fewer wire breaks means increased production through less downtime.
2) With uniform flushing, finishes will improve and be much more consistent as there are no "hot spots" to burn or discolor the part surface.
3) Efficient flushing replenishes the gap with "fresh" dielectric quicker, reducing the recovery time required for reionization. This will allow a reduction of the off-time, which increases the duty cycle and overall efficiency, and is translated into faster cutting speeds. In this case, skimming speed using the flushing fixture was increased 33% over skimming without it.
4) Because the flow of water comes from behind the cut (pushing the debris ahead of the cut), the dielectric in the cut is much cleaner and almost elimi-

nates the problems associated with secondary discharge that can slow the cut down and decrease accuracy due to "bow" or "belly" in the part.

This job ran more quickly and accurately because careful planning showed that enhanced flushing would be mandatory to meet blueprint and time constraints.

CONCLUSION

In situations like this one, it becomes clear that information from builders and tooling companies alone will sometimes no longer be adequate. This is in no way a reflection of a lack of response or concern on the part of the builders of machinery and tooling — it is merely a reminder that an increasing number of shops are not content to pursue the same routine, everyday jobs as everyone else. Therefore, those who are doing business "out-on-the-edge" must continually explore and experiment to *extend* the edge a little further — to "expand the envelope."

Despite the risks involved, fortunately for us, this is the nature of man. Without this quality, Columbus would never have sailed West to prove that the earth was round, nor would Wilber and Orville have dared to "loose the surley bonds of earth…." and prove that manned flight was possible.

The point I am trying to make is that as far as wire EDM goes, sometimes we think we are exploiting this wonderful technology to the fullest extent possible, but that some of the best — the *most significant* improvements in speeds, feeds, accuracy, and reliability — are also sometimes the simplest. Since power supply or computer development are beyond the capabilities of most of us, perhaps the most accessible area of improvement is flushing.

SUMMARY

Those of us involved with EDM are lucky. Lucky in the sense that we can *all* be "pioneers" in pursuing or developing new or more efficient manufacturing technologies. Since we are pioneers, there are very few "rules" other than the basics of theory that have already been established. We have the ability to improvise, to improve upon results of the past by continuing to explore new methods, new procedures, and new attitudes.

Since flushing is so *elementary* yet so *crucial* to EDM performance — wire or sinker — it seems only logical that those who are in command of it will, in the long run, be far more perceptive, productive, and profitable.

Chapter 13

WATER QUALITY

INTRODUCTION

One of the most overlooked and taken-for-granted facets of successful wire-cutting is the water that is used as the dielectric, despite the fact that it is integral to the EDM process. On the surface, this neglect can be easy to understand because, after all, "it's only water," right?

Water is cheap, plentiful, and available at the turn of a valve. So what's the big deal? The big deal, at least for wire-cut EDM, is that without water in the correct quality, quantity, cleanliness, and temperature, *you don't cut*. At least not very well, and perhaps not at all.

WATER QUALITY — FILTRATION

Technically, at least for wire EDM, water quality or cleanliness can be classified into two parts: *mechanical* and *chemical*. First, we will examine how we *mechanically* clean our liquid medium.

Webster's Dictionary defines "mechanical" as:

1) having to do with machinery or tools, and
2) produced by machinery or tools.

Both of these definitions are applicable as there are pumps (machinery) to force the water through the filters (tools) to *mechanically* clean the water. On the surface, this sounds simple — until we examine the many different ways of cleaning our water dielectric.

DISPOSABLE PAPER MEDIA

Disposable paper filters are the most common method of filtration for wire-cut EDM, with filters ranging quite a bit in size, but all sharing a similar design and

173

construction. Almost all disposable paper media cartridge systems are constructed with a central metal tube running through the cartridge. Around this tube is a pleated paper filter medium that typically has a filtration range of 5 to 25 microns. This paper media is sometimes surrounded by a perforated metal jacket or "shell" or a woven-fabric "sock." This assembly is then glued — top and bottom — to the inside of fitted metal end-caps.

Some manufacturers force the water through the filter from the inside to the outside. Because this tends to "inflate" or expand the filter, this limits the amount of pressure that can be used; otherwise the filter can rupture, tear, or burst.

Other manufacturers force the water through the filter from the outside. This type filter is usually contained within a pressure vessel and can withstand much higher pressures than the previous example because it is being compressed instead of expanded.

A great deal of study has been done by OEM's (Original Equipment Manufacturers) and aftermarket suppliers to determine the best pleat shape, pleat size, pleat angle, the direction of water flow, etc. All information gleaned from this research, including the failures and disappointments, contributes to the ongoing pursuit and production of better, more efficient, or longer lasting filter products.

PERMANENT PAPER MEDIA

This type differs from the disposable filter systems in both design, micron size capability, and permanence. One major brand utilizes die-cut cardboard disks or wafers that have been stacked atop one another and tightly compressed. This is called "edge-filtering" as the water is forced through the edges of this stacked media. This product will typically filter out suspended particles down to the 1 micron range.

Instead of discarding the filters when contaminated, they can be "backflushed." This is done by forcing compressed air backwards through the filters which forces the coating of particles to be blown off the outside of the filter media and they are then collected in a waste sump. This procedure, done regularly, will keep filtration pressures within a safe range, and it eliminates messy filter changes.

PLASTIC CARTRIDGE FILTERS

These are typically fine-mesh polyester filter cartridges that are arranged in clusters to filter particles to 5 microns. This type of system makes it possible to remove the cluster from the pressure vessel and wash off the sludge with a garden hose. A second cluster is typically purchased for rapid switchover for minimum downtime. The dirty cluster can then be cleaned while the machine is operating. (*Note:* The sludge material coating these filter clusters may contain harmful or toxic materials. Check with local, state, and federal authorities before allowing effluent to enter public drains or sewers.)

DIATOMACEOUS EARTH FILTERS

This filter medium has been borrowed from the swimming pool industry. These types of filters work by forcing the dirty or contaminated water through a network of filter structures charged or coated with diatomaceous earth (DE). DE is actually the calcified or fossilized bodies of microscopic sea creatures called diatoms. When a filter medium is precoated with a layer of these substances, filtration in the range of 1 to 3 microns can be expected.

The filter element that is to be precoated is made up of specially woven plastic or stainless steel tubes. During the precoating process, the DE is forced into all of the cracks and corners of the element until it is entirely coated and has no openings larger than a few microns. This "clogging" of the larger openings of the steel or plastic braids is what gives this type of system its 3 micron or less filtering capability.

Systems that use DE as a precoat medium can be backflushed by forcing fluid back through the tubes. Since they are flexible, they can expand and distort slightly without damaging the filter. The combination of reversed flow direction and the expansion of the woven braid dislodges the sludge. The sludge then settles into the base of the vessel where it can be removed and disposed of.

ELECTROSTATIC

While still uncommon as far as EDM systems are concerned, it is only fair to include this method as it does have merit, although it provides only half an answer. It still needs to be augmented with a separation device.

Electrostatic separation is a process whereby conductive and semi-conductive particles are "excited" by an electrical charge. This causes them to be attracted to each other, forming larger particles, allowing the use of larger mesh filters or faster centrifuge separation. While this makes their removal easier, further processing is still required.

At present, the single greatest barrier to the acceptance of this type of system is the prohibitive system cost in relation to the cost of the EDM. These systems, while effective in their actual operation, are better suited for other applications rather than EDM, as cost justification is difficult at this time. Another drawback is the requirement and expense of an additional (or existing) system for separation or filtration. Electrostatic separation, at least for EDM, is not a "stand-alone" system. It must be augmented with a filtration or separation system or device.

CENTRIFUGE

This is a unique device to separate and suspend solids from almost any liquid medium. Its use is more common in other industries such as reclaiming washing solutions, cleaning solvents, cutting and grinding oils, silver reclamation, etc.

A centrifuge spins contaminated liquid around in a high-speed, rotating drum that generates literally *tons* of centrifugal force to sling the heavier particulant from the liquid, pressing it onto the inside of the rotating drum. The resulting "g-forces" on the inside surface of the drum can rise to over 3,000 g's (3,000 times one earth gravity). This high-gravity environment causes any suspended particulant that has more mass than its surrounding liquid to be forced against the inside wall of the spinning container. A centrifuge specifically slated for EDM use can easily remove suspended solids below 5 microns in size.

Using a centrifuge-type separation unit can also reduce the problems encountered with wet-filter waste disposal. The particles that are left are compressed into a very dense "cake" that, in most cases, is dry enough to meet most EPA requirements concerning disposal in public landfills. (*Note:* All materials need an approved laboratory analysis to confirm material content for toxicity, leachability, and acceptable disposal method.)

CENTRAL FILTRATION

These are usually "made-to-order" systems that are custom-fit to a plant's specific needs. Many aspects of the user's operation must be taken into consideration before implementation, such as: the number of machines within the system, type of work, workpiece material, type of wire, speed of cut, duration of constant operation, location of machines versus filtration device, plumbing, pumps, inlets, outlets, waste disposal, etc. All of this information is critical to the salesperson's and/ or engineer's recommendations.

When implementing such a system, follow the manufacturer's instructions for operation and maintenance *explicitly*. These systems are typically very reliable, but if they are not maintained according to instructions, you will be faced with the prospect of not just one machine being down but *all* of your machines being down. Not a nice thought.

WATER QUALITY — DEIONIZATION

So far, we have examined a few ways we can mechanically clean the water. So what else could we mean by water quality? Perhaps water *quality* should be defined further. It is usually referred to as cleanliness. For example, "When cutting carbide, your water should be super-clean."

Does this mean that we must use *super*-filters or a *super*-centrifuge? Not at all. This type of water "cleanliness" refers to the *chemical* purity of the water. We have removed the "rock" from the water, now we must remove the minerals that make water conductive. After we treat the water, there should be nothing in the dielectric reservoir except a large quantity of two parts hydrogen per one part oxygen — H_2O — pure water. Pure water does not conduct electricity; it is the elements and compounds that are dissolved within the water that contribute to its conductivity.

To borrow a phrase used in my high-school science class: "Water is the universal solvent," meaning it can eventually dissolve almost anything. Sugar, sand, rock, or steel — all will ultimately succumb to the solvent effects of water — some faster than others, obviously, but all will eventually be changed. If there is any doubt of water's solvency, peek into the huge underground world of Carlsbad Caverns or consider the never-ending cycle of painting and repainting the Golden Gate Bridge that is required to protect it from water's corrosive nature. Many of the materials that water dissolves can be held in solution. And although our supply of tap water has been "purified," it is far from pure. Let's examine this a little closer.

Tap water can contain high levels of minerals depending on geographic area, and these minerals are responsible for what we commonly refer to as "hard" or "soft" water. Minerals that have been dissolved in water and are held in solution are referred to as Total Dissolved Solids (TDS), and are measured in parts per million (ppm).

Water can be contaminated in two different ways. Using the old comparison of sand and sugar, "suspended" solids would be particles of sand or dirt that are mixed with water. Sand and dirt may cloud the water, but these particles can be physically removed by filtration, electrostatic separation, or centrifuge. Sugar, on the other hand, will dissolve in water and it cannot be filtered out but must be removed chemically.

As described earlier, filters remove "suspended" solids from water mechanically, physically filtering out particulants such as metal "chips" and carbon compounds from the wire and workpiece. Mixed-bed resins are used to remove any "dissolved" solids from the water by chemically cleaning or removing metallic ions and minerals from the water.

ION EXCHANGE

So exactly how does my resin system work? By actual definition, ion exchange is "the process of reversible exchange of one ion for another between a solution and ionizable solids, with no substantial change in the structure of the solids."

OK, I'll explain it another way, but it will take longer. After filtration, the water in your wire EDM is recycled over and over again in a carefully designed and balanced deionization system. The conductivity of the water is monitored by a sensor mounted in the dielectric tank and acts as a conductivity "thermostat." It measures how much the water resists electric current (specific resistance), or how much electric current the water can transmit (specific conductance).

By this measure (resistance or conductance) and depending on its setting, the conductivity meter will open or close a valve that regulates the flow of water through a container filled with resin. This resin is in the form of small, plastic beads that have approximately 65% of their ion exchange surfaces on the inside. *Important:* Since so much of the surface area is inside these beads, it is imperative to maintain clean, filtered water (to at least 5 microns), or these small capillaries will quickly become clogged with suspended particles and be rendered useless.

Resin used for EDM is frequently advertised as "nuclear resin" or "nuclear grade." This is a grade or standard that has been set up specifically by the nuclear

Figure 13.1   Mixed-bed resin beads used to deionize water dielectric. (Courtesy of AmeriWater.)

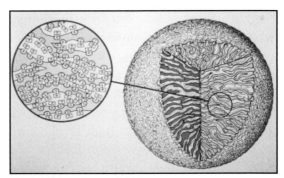

Figure 13.2   Section view of polyester resin bead. Capillaries contain approximately 65% of the bead's surface area. (Courtesy of AmeriWater.)

power industry, one of the main consumers of deionizing resins. (This water is used for cooling in nuclear power generating stations.) They typically require tighter manufacturer's specifications for the resin in respect to particle size, distribution, and organic leachables to provide the highest water purity possible.

These beads have been specially treated with a strong acid for cation resin and a caustic for anion resin. For EDM, the correct mixture or recipe is a ratio of 60% anion to 40% cation resin, hence the term "mixed-bed resin."

Positive Ions	Negative Ions
Aluminum	Carbonate
Calcium	Bicarbonate
Magnesium	Sulfate
Sodium	Chloride
Potassium	Nitrate
Ferrous	Phosphate

Figure 13.3   Tap water commonly contains a variety of minerals.

All dissolved solids in water have either positively charges atoms (cations) or negatively charged atoms (anions). (See Figure 13.3.) When these minerals are passed through the resin system, the cations are replaced by hydrogen (H) and anions are replaced by hydroxide (OH). Eighth-grade science tells us that: H + OH = H_2O, or pure, nonconductive water.

One final issue, but far from the least important, is *waste disposal*. Even before use, mixed-bed resin is already highly toxic to the environment due to its strong acid (cation resin) and base (anion resin) content. In addition, wire EDM is commonly used to cut tool steels which contain many elements that are classified as harmful to the environment — chromium, manganese, nickel, and vanadium, to name a few. Also identified as toxic in OSHA's Code of Federal Regulations (CFR) are copper and zinc — *the very components of brass wire!* So, without exception, after use, resin will not only possess the initial toxic properties of acids and bases, but it will also likely contain high levels of heavy metals and/or other toxic compounds, so its disposal can be a problem. Either go through a supply source that regenerates and/or disposes spent resin as a service to their customers, or check with the proper authorities in your area to determine the correct handling, transportation, and disposal procedures.

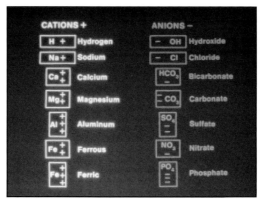

Figure 13.4   Schematic of ion exchange during ionization process. (Courtesy of AmeriWater.)

TEMPERATURE

Now that our water has been filtered and deionized, what next? Can we use it now? Actually yes, but not for long. In a very short time, the water temperature will rise and inhibit cutting speed and accuracy. We must cool the water before reusing it.

Although the temperature of the EDM spark is very high, it is not the main reason for the rapid rise in water temperature — the pumps are. Ironically, the most integral components of the dielectric system contribute most to affecting speed and accuracy. Today's modern equipment, with flushing pressures to 300 psi, require efficient, high-output water pumps. The motors that drive these pumps typically have a temperature rise of at least 38°C (100°F) or more. This heat is then transferred to the water from the pump parts that are naturally exposed to the water. This is the heat that must be removed from the water with either a cooling tower or refrigeration unit.

A large share of part accuracy (besides the capability to cool the machine) is the result of maintaining uniform machine component, tooling, and part temperatures in relation to the temperature of the room. Temperature changes during cutting can cause the part to bend, twist, or distort in addition to "growing" or elongating due to the coefficient of expansion for a given material.

For a reference or benchmark, one of the most common tool steels in use is AISI D2. This is a very stable, high carbon, high chromium tool steel with extremely high wear resistance. It is known to be practically free from size change after proper

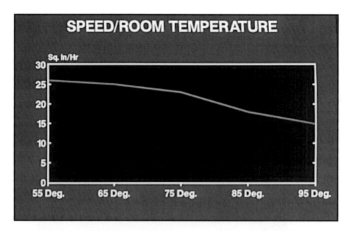

Figure 13.5   Chart demonstrating loss of speed as temperatures increase. (Courtesy of MC Machinery, Inc.)

treatment. Even as stable as this material is, a user can expect a 0.00006" per inch error for every 10° F rise over room temperature.

For example, if a 10" long part is 16° warmer than ambient room temperature, it will grow about 0.001" in length. If cut at this temperature, when this part has cooled to room temperature, it will shrink and measure 0.001" smaller than desired. Without a cooler or chiller, wire-cut EDM's can have a 40° to 50°F rise in water temperature very quickly. The water temperature can rise from stone-cold in the morning (60°) to warm to the touch (100°+), in just a few hours. If we cut the same 10" long part in 100° water, it will be 0.0025" undersize at room temperature.

For accuracy's sake, the temperature of the water should be as close to room temperature as possible.

Speed is the other factor influenced by water temperature. It is already heavily documented that cutting speeds fall off rapidly as the water temperature rises.

For example, a machine cutting at 25 sq. in./hr. at room temperature (72°) will slow to 22 sq. in./hr. at 80°, a reduction of almost 14%. If allowed to climb to over 100° as in the previous example, cutting speeds of even 10 sq. in./hr. would be difficult to obtain.

Obversely, cutting speeds tend to increase as the water temperature is reduced. For reference purposes, our 25 sq. in./hr. machine above might cut 27 or 28 square inches at 50°. According to tests conducted on temperature effects on cutting speeds, this trend will continue until water temperatures reach 36° to 38°, then cutting speeds will level off.

Note: Don't run off and start buying "super-chillers" for your wire machines expecting to cut faster than your competition. The major problem with using significantly lower cutting temperatures to gain speed is the accompanying reduction in part accuracies. Even if one compensates for part growth with offsets, except in the simplest parts, who can predict exactly how the parts will grow? Will growth be uniform? What about thick or thin sections? What about hole locations? Also, every material has a different coefficient of expansion. The group that conducted and documented these tests on the effects of lowered temperatures strongly cautions against the pursuit of higher cutting speeds through reduced temperatures, and I agree. Don't feel like you might be missing something. If this method was controllable, everybody would be doing it by now!

SUMMARY

"Flushing, flushing, and flushing" is just as important in wire-cut operations as it is for vertical or sinker operations. How we administer it was covered in the last chapter. What we are flushing *with* is just as important as how we present it. To successfully cut at optimum speeds, while maintaining part accuracy, we must do the following to our water.

1) **Filtration**. Filters should have the finest mesh economically feasible for the job. This will allow faster cutting because of clean cutting conditions, which means less secondary discharge. Fine-mesh filters will also ensure longer, more efficient life from resin.

2) **Deionization**. The resin system is an integral portion of the wire-cut machine. We can get away with unfiltered water for a short time, but once the deionization process gets past a certain point, cutting will stop as quickly as if you turned the machine off.

3) **Temperature Control**. It is a matter of documentation and fact that a wire-cut EDM will cut parts faster and produce more consistently accurate parts with a temperature-controlled dielectric system than without one.

Chapter 14

CARBIDE AND
THE EDM PROCESS

INTRODUCTION

Tungsten carbide is an extremely hard material used extensively in manufacturing because of its superior wear and corrosion resistance. Second in hardness only to diamonds, other carbide compounds, and aluminum oxide, its production and machining present some of the most difficult challenges in manufacturing today. With the exception of diamond-charged grinding wheels, EDM is the only economical method of machining this extremely hard material.

Used extensively throughout manufacturing, tungsten carbide is most commonly used in cemented and solid carbide cutting tools, stamping, forming and compacting dies, mechanical wear-parts, and electrical contacts. There are many other types of carbide but for ease of discussion, the term "carbide" will be used throughout the rest of this chapter.

While EDM is one of the most common methods used to machine carbides, there are some "built-in" difficulties related to this process. The unavoidable recast layer left by the EDM process is one, while thermal and stress-related micro-cracking are others. Further, as the demands for better surface finishes and longer tool life of carbide components continue to increase, the effects of the electrolytic damage that can occur during the wire-cut process has to be better understood and prevented. Generally speaking, wire EDM can affect carbides more than vertical EDM. The reasons for this will be explained later within this chapter, but to better understand the difficulties in EDMing carbide we must first examine the composition of carbides themselves.

THE MAKING OF CARBIDE

Powder metallurgy is the technology used for production of tungsten carbide. Tungsten's extremely high melting point, in excess of 3370°C (6098°F), prevents its production by standard foundry casting methods. Instead, it is mixed in powder-form with other metal powders or "binder materials" and then vacuum or hydrogen-sintered. First, let's examine the chemistry of carbide, which is relatively simple. It's the *processing* of this chemistry that drives the art of carbide making.

Tungsten is obtained from Wolframite, the ore of the element Wolfram. In preparation for making tungsten carbide, it is reduced to a fine powder in a hammer or ball mill before being sifted, screened, and graded. Carbide powder is produced by heating a mixture of 94 parts tungsten powder and 6 parts carbon to approximately 1537°C (2800°F), to carburize this mixture into tungsten carbide. Although this process gives you tungsten carbide, it is not useable in this form. Even after pressing and sintering, it can not be used in any traditional carbide application. Before it can be used commercially, a binding agent, usually cobalt, must be added for strength.

Figure 14.1   View of coarse-grain carbide shows sintered structure. Dark particles are tungsten, light area is cobalt matrix. (Courtesy of Carbidie, Inc.)

Cobalt is a very hard, conductive metal and it is the most common matrix material for cemented carbides. Cobalt powder, in amounts ranging from 3 to 30% by weight, is ball-milled together with tungsten carbide for several days, forming a very intimate blend. A higher percentage of cobalt will increase the material's strength while reducing it's hardness. Decreasing the cobalt content has the opposite effect.

Next, a small amount of paraffin is added to the carbide/cobalt powder mix to help hold this mixture together when it is pressed into a billet or shape, and to act as a lubricant to aid pressing and reduce wear of the compacting dies. This is called "waxing" and, after this step, the mixture is ready for the next step: pressing or compacting.

COMPACTION

Compaction is the process that presses or molds the carbide powder into shapes under high pressures. These carbide billets, blanks, and pre-forms can be made in three ways:

1) **Hydrostatic pressing**. This method places the raw carbide/cobalt-powder mixture inside a tough rubber bladder within a steel mold that is then sealed inside a pressure vessel, usually containing water. Under high pressure, normally between 25,000 and 30,000 psi, the rubber bladder compresses the powder mixture into square or round billets or briquettes. These billets are made large enough so they can later be saw-cut or fabricated into numerous smaller shapes before undergoing any sintering processes.

2) **Hydraulic pressing**. If large production quantities of a carbide shape or part are needed, they can be made in compacting dies which are themselves made out of carbide. A compacting die is a mold or impression of the part that, after being filled with the correct powder grade, is compacted at 30,000 psi. These pre-forms or near-net shapes are called "compacts." The common "throw-away" or disposable carbide insert used for milling or turning is a good example of a production part made in compacting dies.

3) **Extrusion**. This method requires that a special polymer be added to the carbide powder to give the powder mixture a more gummy texture. In this composition it can then be extruded under pressure through a round opening in a long continuous shape (much like toothpaste). Blanks for carbide drills, reamers, and end mills are produced this way.

Regardless of the method, carbide compaction can approach a maximum density of only 58%. At this point, the pre-forms are strong enough to be easily handled and prepared for further processing or "green-machining."

SINTERING

Although it is not always necessary, this "green" blank (or compact) can be half-sintered or pre-sintered. After pre-sintering, the strength and hardness of the compact will approximate blackboard chalk which is sufficient enough to allow it to be green-machined to oversized dimensions using carbide or diamond-tipped cutting tools. It is important to note here that all dimensions of a compact must allow for the substantial shrinkage that will occur during final sintering. Because of its high porosity at this stage, carbide compacts will shrink about 22% in length or about 40% volumetrically during final sintering, depending upon the grain size, cobalt content and part thickness.

The final sintering temperature is around 1500°C (2732°F) and will completely melt the cobalt. In liquid phase, the cobalt flows through the open tungsten latticework and surrounds the tungsten particles. When cooled, the cobalt matrix becomes the binder material that holds the tungsten particles in place. This is now a sintered product, not an alloy. During final sintering the part will achieve almost

Figure 14.2   Before-and-after view displays the extreme shrinkage carbide undergoes during sintering. (Courtesy of Carbidie, Inc.)

100% density and, surprisingly, even though the compact will shrink 40% in volume, *it will weigh the same as it did before final sintering*. This fact is important. It is related to problems we will likely encounter later.

Modern carbide-makers subscribe to "traditional recipes" for "traditional grades." Even though the metallurgy or mixtures of powder weights are the same from grade-to-grade, controversy arises when discussing the new "EDM recipes" requiring specific heating, "soaking," and cooling "ramp-rates." These are considered proprietary methods and are jealously guarded by their owners, so we will not/ cannot examine them here.

At this point in its production, our carbide product can now be inspected, inventoried and supplied as blanks for industry. It can be sold in its present billet form or be cut and ground to customer specifications.

POST-PROCESSING

There are certain methods and processes that reportedly improve the quality of the carbide itself, and another that actually determines the quality of the carbide after processing. The pros and cons of certain post-processing methods are no less controversial than those of carbide making. The following examples are both embraced and rejected by different carbide manufacturers.

HIPing. When producing larger sized compacts with more mass or heavy cross-sections, shrinkage can be so severe it can cause internal voids and cracking. Hot isostatic pressing (or "HIPing") has been used for years to reduce or eliminate this condition. HIPing is the process of placing the compact within a thick-walled pressure vessel and reheating it to a temperature just below the melting temperature of cobalt. While in this "plastic" state, high pressures are applied, usually via a compressed, inert gas such as argon. This method has been used to compress and eliminate any internal voids and "knit" together any possible micro cracks. For years it has been claimed that this method improves material strength and provides a more dense, homogeneous composition, but recently there is a growing number of carbide industry professionals who are questioning whether HIPing actually provides these benefits. Some engineers maintain that if the billet is prepared correctly beforehand, it will not need any further processing, including HIPing. In stark contrast (especially financially), some mills HIP practically every billet they produce while others HIP only upon a customer's request. One manufacturer suggests that processing technology has improved powders and processes to the degree that HIPing is no longer necessary. One can easily see there is no small controversy here.

Cryogenics. No less controversial, the verdict is still out regarding cryogenic treatment. It involves the sub-zero freezing of the compact to change or improve its grain structure. Once hoped to be "the cure" for all carbide woes, interest in this area has lessened and the industry as a whole seems disenchanted with this process, at least for the time being.

Ultrasonics. This procedure is showing great promise in the testing and elimination of stresses in many materials, carbide among them. Ultrasonic soundwave technology is not actually used to *make* a batch of carbide. Instead, it is used as a method of detecting stress and stress-relieving a material, thereby helping to ensure it's quality and integrity prior to sale. It is even used to check workpiece characteristics in-between machining operations.

The procedure is as follows: varying frequencies of intersecting ultrasonic sound waves are applied to a part or sample until a certain resonance is detected on a video screen. The frequencies used will vary depending upon the material and shape of the object being tested. After first detecting the stressed area, the sound waves are adjusted to cause high-frequency vibrations that cause internal friction and heat (not unlike the way organic materials react in a microwave oven). This process is reported to "relax" and eliminate internal stresses in many materials, carbide included. Ultrasonics are not limited to carbide or the world of EDM. This process is reported to be so responsive in certain types of materials that it is being applied to eliminate the stresses inherent in the heat affected zone (HAZ) of rapidly cooling weldments.

At this writing, interest in ultrasonics is increasing and showing promise, but, like other technology within the carbide industry regarding carbide making and testing, the technique continues to have its skeptics.

EDM AND CARBIDE

Having a better understanding of the making of carbide will aid us in understanding its behavior when it is EDMed. Let's examine what happens.

The cobalt matrix that holds the tungsten particles in place has greater conductivity than tungsten. In addition, there is a large difference in their melting temperatures; Cobalt melts at approximately 1480°C (2696°F) while tungsten melts at approximately 3370°C (6098°F). The EDM spark tends to flow around the "colder" (both in conductivity and in melting temperature) tungsten particles and strike the less resilient cobalt binder. This can leave a surface of partially melted tungsten particles bound loosely by an insufficient amount of cobalt binder. These exposed tungsten particles can flake off, changing the size and shape of the detail and lead to premature part or tool failure.

This degradation of the EDMed surface is aggravated even further by wire EDM because deionized water is the typical dielectric and it supports electrolysis and oxidation which will further leach away any exposed cobalt and contribute to the already undesirable condition of cobalt depletion. More on electrolysis later.

When EDMing carbide, the most efficient part of the EDM cycle is the initial impact of the spark which results in vaporization. Because carbides are so susceptible to thermal damage, we will *always* use short on-times (high frequencies) to reduce the possibility of melting versus vaporization. Longer on-times allow current to flow longer, producing larger amounts of heat. This allows for faster machining, but it will also produce thicker layers of recast material. The recast layer consists of partially melted nodules of tungsten held loosely in place by a severely depleted and structurally weak surface of cobalt. The deeper this layer is, the greater the chance for cracking caused by thermal stresses on the EDMed surface.

Remember that many carbides are predisposed to cracking due to the high internal stresses that naturally develop when the cooling carbide/cobalt matrix shrinks 40% volumetrically during sintering. In addition, with all of the variables inherent in EDMing carbides, including electrolysis and corrosion, it's no wonder we often encounter difficulties.

STRESS

Figure 14.3 shows a very simplified example of the X-Y directions of residual stresses around a tungsten particle. The graphic displays stress forces of the cobalt matrix pulling at the tungsten particle in only four directions and in only one plane. Now consider the amount of stress in the actual part when these forces are pulling in *all directions* and in *all planes* radially around each particle. Perhaps you can now begin to appreciate the amount of inherent stress resident within a carbide blank.

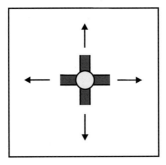

Figure 14.3   Omnidirectional stress.

Most of us are familiar with the behavior of a cold-rolled steel plate that has had more material removed from one side than the other. It has an "elastic skin" (not actually, but it makes this discussion much easier), and when the "elastic" is removed from one side, the other side can pull the part out of shape.

There is a great deal of this type of stress or "elasticity" caused by shrinkage throughout the entire carbide blank. A good example of this is the old "balls and bands" analogy. The "balls" in this case are tungsten particles, and the "bands" are cobalt. During the sintering process, the cobalt melts and cools around the tungsten "balls." It shrinks in all directions and exerts a tremendous amount of tension that holds the tungsten "balls" in place. If some of the cobalt "bands" are cut, the tension in that direction is relaxed, while the ones still attached will retract or pull the ball away from the cut. As the cut proceeds and more and more stress is relieved behind the cut, the opposite end can no longer withstand the built-up forces and stress cracking will occur, typically in front of the wire path (see Figure 14.4).

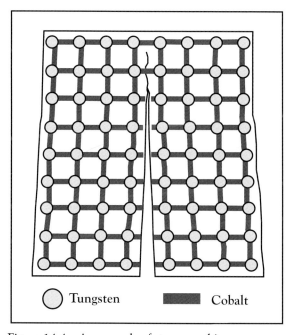

Figure 14.4   An example of stress cracking.

It was for this reason that, until recently, it was recommended that carbide blanks should be ordered as close to the finished size as is practical. Shops were advised *not* to buy large pieces of carbide and cut small ones from it, as is commonly done with ferrous tool steels. This was before new carbide technology and "carbides for EDM."

Traditionally, large carbide parts should have strategically placed "stress-cuts" made where applicable. These can be saw-cut before sintering, as the blank material is soft and can easily be cut with diamond saws. These same type of stress relieving cuts can be made by the wire EDM, but that method is much slower in comparison, and therefore more expensive. These "stress-cuts" will allow the part to gradually "relax" as stress is relieved from each area, preventing catastrophic stress cracking. Good part design along with "strategic" EDMing can play a large factor in preventing these cracks from occurring but the time and costs to do these cuts must be considered in the cost/evaluation process.

TENSION CRACKING

If the stress already resident within the carbide blank itself doesn't present enough problems, the EDM process compounds it by adding additional stress to the material which results in what is called tension cracking. Tension cracking in EDMed carbide is usually caused by a thick layer of recast material that occurs when low frequencies (long spark duration) elevate the temperatures of the gap area. This problem is encountered more often in sinker operations because the oil dielectric acts as an insulator, retaining heat instead of a fast quench like a water dielectric can provide.

Some materials can sink heat rapidly. That is, the surrounding material can draw or transfer heat away from the spark area. Carbide is not one of these materials. This poor transfer of heat keeps most of the spark temperature localized within a very thin layer. This area expands rapidly as it is melted, and when the spark is turned off, some of this molten material is drawn back to the parent material by surface tension and cooling effects. When this layer of recast material cools, it shrinks towards the center of each individual crater, imparting a considerable amount of stress on itself and the adjoining areas. With every crater playing "tug-of-war" with all of the other craters, this highly stressed "skin" can crack from temperature changes or mechanical shock. In order to keep this "skin" as thin as possible, low power and high frequencies should be used.

EFFECTS OF DIELECTRICS ON CARBIDE

As different as oil and water are, they will also contribute to a marked difference in the surface integrity of a workpiece material. Carbide will be no exception.

OIL

Parts EDMed with oil dielectrics will not experience any electrolytic damage at all but instead will become more susceptible to thermal damage and microcracking. This is due primarily to the heat insulating effects of the oil and the carbon rich environment of the spark gap. The high concentrations of carbon trapped within the remelted material can make the EDMed surface much harder and more brittle than a water dielectric will. Therefore, the potential for thermal damage is much higher when using oil for a dielectric.

There are specialty wire machines available that can wire-cut parts submerged in oil. This completely eliminates the water caused problems of electrolysis, rust

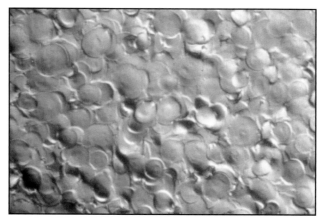

Figure 14.5   Surface of carbide EDMed in oil dielectric.
Melting gives craters a "puddle-like" appearance.

and corrosion. For the right application, this type of machine can be ideal. Presently, these machines use a dc generator and provide good finishes and a reported 30% increase in die life when compared to parts produced with a similar dc generator using deionized water.

Note: While an increase of up to 30% more die life is certainly desirable, it does not come without significant concessions. The trade-off for this increased die life is substantially slower cutting speeds. Using 0.008" diameter brass wire, speeds for wire-cutting a one-half inch thick C-4 carbide workpiece submerged in oil are about 5 sq. in./hr. for a full-width roughing cut. Skim-cut speeds run about 8 to 10 sq. in./hr., depending upon the finish desired. Careful evaluations of cutting times and job costs must be made when considering an oil-only machine, usually dedicated to specific applications. If the limited cutting speed of cutting in oil is a negative, machines using oil *and* water are also available. A user can rough the part flushing with water for machining speed, then change over to oil for the finishing skims. Selection of either type of machine must be driven by the application.

WATER

Some builders have developed new generators using water dielectrics and zero-electrolysis circuitry and claim superior results with little sacrifice in speed. Some of these models of wire machines are available with ac generators or circuits where current is switchable from dc to ac. The dc current would be used for speed and roughing cuts then switched to ac current for finishing. Using ac generators should reduce electrolytic damage and corrosion by about 50%, but it will also reduce cutting speeds by approximately the same amount. To compensate for this, machine builders have added special circuitry to ac generators in the form of patented "black boxes" or phased-wave generators to push cutting speeds to nearly that of dc generators but without the resultant part damage typically found when using dc current and water dielectrics. (As of this writing, no information is available concerning the performance of an ac generator using *oil* as a dielectric.)

Parts that have been EDMed in oil will have a recast layer that is martensetic, or high in carbon. This layer will be harder and more brittle than the parent material because of the assimilation of carbon atoms into the molten workpiece material, essentially being carburized. The recast layer that is left by water dielectrics is usually slightly softer due to oxidation. This occurs as the hydrogen and oxygen are driven from the water. Hydrogen embrittlement may also occur in certain alloys but this condition can be corrected by properly "baking" the part to drive off any hydrogen remaining in the EDMed surface.

Figure 14.6   Surface of carbide EDMed with water. Water's faster quenching gives the surface a granular appearance.

WHY OIL? WHY WATER?

In any case, sinker or wire, all cuts will be much faster using water as a dielectric. Why? Water is a much better flushing medium. It is "cooler" in many respects and obviously less viscous than oil. Instead of insulating and retaining heat in the spark gap, water dielectrics not only sink heat from the spark gap much better (it is almost always cooled or refrigerated), but also flush better. This is because water is naturally thinner than oil and possesses what is referred to as better "gap penetration." This is the speed and ability at which a liquid can naturally move within a confined space. In EDM's case, of course, there are extremely small spark gaps. Mirror finishing settings are so low and spark gaps are so small that even capillary action gap penetration is welcomed and water does this much faster and better than oil.

If machining speeds are always faster using water as a dielectric, then why do sinkers use oil? Primarily to reduce electrode wear.

Years ago, "standard" polarity was a negative electrode. This provided excellent machining speeds but at great sacrifice of the electrode because of high electrode wear. Since the design, fabrication and flushing of an electrode is often more costly than the actual EDM operation, it is very desirable to find ways to protect the

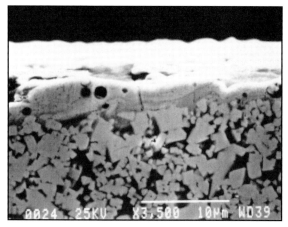

Figure 14.7   Section view of carbide surface EDMed in oil. Note severe example of recast layer, including gas bubbles. Part already displays extensive microcracking. (Courtesy of MC Machinery Systems, Inc.)

electrode from excessive wear. This can be achieved by using positive polarity electrodes. Even though positive polarity electrodes slow machining, their use reduces electrode wear from as much as 50%, to 1% or less. Today, in favor of protecting the electrode, positive polarity is considered "standard."

Experimentation with water dielectric in sinkers began more in trying to eliminate oil contamination of certain types of parts than as a quest for machining speed. In this manner it was learned that cutting speeds in sinker operations can be increased up to five times by using a water dielectric like the wire EDM uses. A comparison of a graphite electrode against a steel workpiece is approximately 3.2 g/min. at 40A for oil, while the same set-up in water will run at 6.0 g/min. at 40A and speed continues to increase with additional amperage. Presently, published tables showing machining speeds for water dielectrics stop at 80 g/min., or 37.6 cu. in./hr. at 320A. Surface finishes will be rougher also. A best finish of 25 RMS ($4\mu R_{max}$) using an oil dielectric will increase to 125 RMS ($20\mu R_{max}$) with water.

Figure 14.8   Section view of carbide surface EDMed in water using anti-electrolysis generator. (Courtesy of MC Machines, Inc.)

The special water based dielectric used in these machines is reported to be quite expensive and, because it boils away so quickly, it must be replenished much more often than EDM oils. Further, to keep up with the high metal removal rates, a cumbersome filtration and dielectric system must be maintained using diatomaceous earth (a known carcinogen), paper-media, and a conveyer system while consuming large amounts of deionizing resin. Unfortunately, the faster machining speeds realized using water in a sinker do not offset the high electrode wear, rougher finishes and high operating and maintenance costs. These negative aspects have made this type of machine cost prohibitive and production machines using water instead of oil have twice been released and withdrawn from the U.S. market.

EDM PITFALLS

We now know that besides the fact that carbides are among the hardest materials in the world, they are also fraught with stresses so extreme that they are almost self-destructive. And that's even before we try to EDM them! Difficult as this can be, there are several methods to insure the successful EDMing of this unique material. Since wire-cutting carbide is the most common EDM operation and especially since wire-cutting presents the most problems encountered in machining, the following solutions are all wire-oriented.

Since the EDM process itself is going to be "hard" on carbides, starting with a quality carbide blank is essential. Quality control throughout the manufacture of carbide is a critical concern of all suppliers, and improvements in powder, blending, sintering and cooling are constantly being made, continually improving the product. The basic carbide blank or compact available from carbide mills today is of very high quality.

In my own experience, if difficulties are encountered during EDMing, or if the carbide part fails prematurely in use, today's carbide product is the last suspect. If EDMing difficulties or premature part failure are encountered, it is usually attributed to one or more factors and seldom caused by the material itself. These are:

1) Selecting the wrong grade of carbide for the application;
2) EDMing a carbide using incorrect cutting strategies;
3) EDMing a carbide using incorrect cutting parameters;
4) EDMing a carbide using the incorrect wire type; and
5) EDMing a carbide using an older machine/generator that *creates* carbide problems.

Avoidance of the above items falls entirely on the user's side of responsibility. One of the few problems that can belong solely to the carbide makers is the occasional impurity or inclusion found within a carbide part. This is typically a very small, non-conductive particle trapped within the carbide itself. Under a microscope these inclusions usually appear glass or slag-like, and they usually result from contamination outside of carbide making such as a small grain of sand or silicon from a worker's clothing. Unfortunately, this particle will not always be found until the part is nearly finished and, therefore, expensive.

The reason that the particles are not always found until skimming is that during a roughing cut, where energy settings are higher, if an inclusion or impurity in the material is encountered, the spark is usually "large enough" to erode away the surrounding binder material that holds it in place, allowing it to be released and flushed away. Its removal is more likened to being "dislodged" rather than being cut away. Under roughing conditions, this is less a problem but if the inclusion is exposed or nearly exposed by the high-powered rough cut, a problem can arise during the subsequent skim cuts. Power settings used in skim or trim cuts are usually too low to fully "dislodge" an impurity, so the wire will either "ride" over it or break against it. If the wire rides over it, it will leave a witness line on the part running from top to bottom of the cut surface that will reveal the height and width of the exposed particle.

Although the problem of impurities within the carbide still occurs occasionally, it appears much less frequently than in the past.

EDM STRATEGIES AND TECHNIQUES

Because the internal structure of carbide is very small, very dense and is easily damaged by prolonged heat, we would always EDM it using reduced amperage and high frequencies (short duration [<10μsec] on- and off-times) regardless of whether it is being machined by sinker or wire. A short duration spark will vaporize a crater but not allow it to continue to melt a larger one. This keeps thermal damage to the heat sensitive carbide workpiece to a minimum.

Vertical EDM

Sinker procedures specify negative polarity at the electrode using either metallic electrodes, preferably copper-tungsten, or a super fine grain graphite, possibly copper-infiltrated. Electrode wear will be high, anywhere from 30 to 100%+ depending upon electrode material and generator settings. Amperage should be used conservatively to prevent too much residual heat build-up, especially during roughing. It is understandably easy to get frustrated or impatient with safe but slow cutting speeds but do not be tempted to approach carbide too aggressively. Capacitance can be added during roughing operations to increase cutting speed but the use of high capacitance will negatively affect the finish and surface integrity if used during final finishing operations.

Wire EDM

Wire strategies could start by selecting a zinc-coated, diffusion-annealed or other specialty wire made expressly for cutting carbide. Different generator ages and origins and the grade of carbide will dictate the ideal type or grade of wire. Check with the builder of your machine for their recommendations before making your selection. Using the proper premium wire will help in obtaining higher cutting speeds, smoother surface finishes and better surface integrity.

If cutting speed is not a large issue, another way to improve surface finish and integrity is to use a smaller diameter wire. A smaller diameter wire has less surface area and, because it has a naturally smaller "zone of influence," lower power settings can be used in skim cuts, contributing to an overall better machined surface.

You must also insure the dielectric's cleanliness, both chemical and mechanical. Because the chemistry of the water and workpiece is taxing to the resin system, resin consumption will increase when cutting carbide. Some paper filters may require precoating because carbide "chips" are so small, many can pass right through the filter until the filter pores are "made smaller" by previously trapped particulants. When wire-cutting carbide, many experienced users will often cut a steel "throw-away" part immediately after changing filters. This fast, powerful cut is performed only to rapidly generate steel "chips," which are larger than carbide "chips" and will precoat a filter faster. Precoating a 5 micron filter in this manner will allow it to filter down to approximately 3 microns, which is a significant reduction. When wire-cutting carbide, filter precision to 1 micron or less is ideal.

MECHANICAL STRESS RELIEVING

There are precautions that can be taken in preparation of the carbide blank to prevent problems of stress cracking or movement when wire-cutting carbide. In every case, extra programming time will be required although sometimes, on contoured relief cuts, the same program can be used but with larger offsets programmed in as is shown in Figure 14.11. All relief cuts should be made a safe distance away from the finished detail. Even though stress-cuts are a cutting strategy with no requirements for accuracy or finish, do not approach these too aggressively or the stress-cuts themselves can cause problems.

SECTIONAL

These are cuts made in a detail to release some of the internal stresses of the blank before cutting the actual part. After roughing out the center hole, cuts are made radially from center (Figure 14.9a), stopping 0.060" to 0.100" from the finished wall. Single cuts within a detail are better than none at all, but multiple cuts, especially if they are angular in relation to the detail, are much better (see Figure 14.9b). In larger details or critical areas, it is sometimes advisable to make these slots opposite each other rather than progressing around the part radially. For example, imagine the workpiece is a clock face. Making stress-cuts first at the 12 o'clock position, then at the opposite side at 6 o'clock, then to 9 o'clock and then back across to 3 o'clock is better than cuts made clockwise at 12, 3, 6 and 9 o'clock.

CONTOUR

Another common practice of mechanical stress relief is to wire-cut contoured

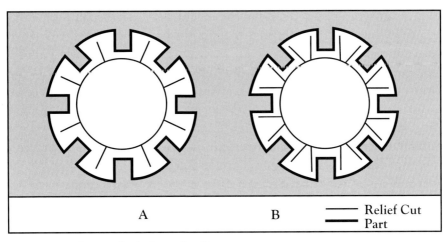

Figure 14.9   Sectional relief cuts for dies.

relief cuts 0.060" to 0.100" away from, but parallel with, the actual contour (see Figure 14.10 for punch or core pin type details). By making the initial contour cut around the actual detail, this will almost wholly "free" the part from the stress-ridden blank prior to finishing. This precaution will allow any stress movement caused by wire-cutting to occur before final finishing.

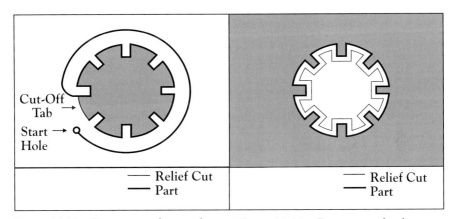

Figure 14.10   Contour cut for punches.   Figure 14.11   Contour cut for dies.

The contour cut shown in Figure 14.11 is effective in reducing internal stresses within a die block. This will take longer (obviously), and will increase operating costs in machine time, wire consumption, etc., but this is a proven method and almost 100% successful in eliminating stress-related cracking (these diagrams have been constructed to demonstrate carbide applications, but can also be used with success in wire-cutting other high-stress materials).

SUMMARY

Today's modern power supplies and controls offer major improvements over older ones. On the machine tool side, finer resolutions of servo systems and the corresponding finer servo response allows the machine to react faster and with greater sensitivity to the low power settings used in finishing operations.

Dielectric systems have improved in both filtration and deionization to provide the "super-clean" water required for fine finishes and surface integrity.

Within the generator, high-speed transistorized circuitry that reduces or eliminates rust, corrosion and electrolytic damage, contributes greatly to the success in EDMing carbide and the improvements of carbide surface integrity. Some machines have eliminated capacitance circuitry entirely, while others have incorporated special circuitry designed especially for carbide machining.

A capacitor is an electrical device used to store electricity. Since water and electricity behave alike in many ways, we will use water as an analogy. For the sake of discussion, a capacitor is like an empty glass. The current can be likened to water running from a hose. When using capacitance, it is like holding the glass under a hose, and filling it up. When the glass is full to the brim, it is rapidly "dumped" out. In reality, when the capacitor reaches its "capacity," it discharges. In effect, the current is "held back" and stored for a short period of time. When it is released, the same amount of volume (current) is delivered, but with a lot more "impact." When the time for this process is compressed to micro- or nanoseconds, and is repeated thousands of times a second, it becomes the equivalent of holding your thumb over the end of the hose. More work can be accomplished without increasing the volume of water (or current), but, unfortunately, this can often damage the finished surface and increase electrode wear.

Manufacturers incorporated capacitors in their power supplies and generators to increase cutting speed but users had to accept the damage it sometimes did to the EDMed surfaces. These highly charged capacitance sparks are not easily controlled and can discharge at random, without the precision of transistor switching. Remember our glass of water analogy? Sometimes the glass would be dumped or discharged before the glass was all the way full. Other times it can fill up "over the rim," contained only by surface tension. Because the amount of stored water (current) can vary, this can cause some EDM craters to be larger or deeper than others. Consequently, high capacitance finishes are not uniform, will have more recast, and tend to retain higher levels of thermal stresses.

Sinkers continue to use capacitance but most wire machines don't unless they are in the form of specialized circuitry especially for EDMing carbide.

The most effective methods used to wire-cut carbide today are low power settings pulsed at very high frequencies with minimal capacitance especially during finishing operations. This provides a very controllable spark, and uniform crater depth with minimal thermal damage done to the EDMed surfaces. Users will have to decide whether cutting carbide submerged in oil or using an antielectrolysis generator is best for their needs.

No one is going to become an expert by reading a chapter in a book or a magazine article, so my suggestion is: If you are still experiencing difficulties EDMing carbide, consult an applications engineer with the builder or importer of your EDM equipment. These people are very knowledgeable of their products and one of their primary functions is to help ensure your success with their equipment.

Another avenue of information and assistance could be from the carbide mill or distributor. Many times their engineers and metallurgists have worked with EDM builders to help develop settings and cutting conditions. Over time, they too, have learned a great deal about EDM and it's effects.

In summary, carbide still remains one of the most challenging materials, but by knowing its properties and its behavior during the EDM process, you have the knowledge that will go a long way to help ensure success.

Chapter 15

MOLDMAKING USING WIRE EDM

INTRODUCTION

Since the introduction and acceptance of EDM in manufacturing, it has been assigned to more or less "typical" roles and disciplines. "Typical" meaning that vertical (or ram) EDM was delegated to moldmaking and mold-type operations, while wire EDM was most commonly associated with the tool and die or stamping industries. More recently, wire EDM has found its way into production and other applications. But perhaps the single fastest growing market for this type of product is in moldmaking. We will examine the reasons for this new interest in a supposedly nonrelated machining method, and its impact on moldmaking, both physically and fiscally.

"MOLDMAKING 101"

Moldmaking is an age-old craft that, in theory, has not changed since the Iron Age: a tool or part is made by pouring or forcing a molten material into a preshaped impression or cavity. Obviously, many refinements have been made throughout the centuries, but the premise of moldmaking still remains the same — creating an impression or cavity from which a part is replicated.

There are many different types of molds and molding processes. Some molded materials are poured, while other materials are injected, inflated, transferred, or expanded by chemical reaction. Other types of molds are not poured or injected but are struck, such as forging dies or coining dies that make everything from the change in your pocket to automobile crankshafts, from the cutlery on your dinner table to steel railroad wheels.

The following list is only a small example of the many different disciplines that require cavity making:

Blow Molds	Compression Molds
Forging Dies	Investment Castings
Plastic Injection	Reaction Molds
Rubber Molds	Transfer Molds
Coining Dies	Die Castings
Form Dies	Metal Injection
Powdered Metal Dies	Rotation Molding
Thermoset Molds	Vacuum Form.

As a rule, except when leaving stock for machining, the closer the shape of a cavity to that of the molded or cast object, the better. Plastic parts are usually molded in their final shape with very few secondary operations required, if any. Except for small, precision die cast parts, very few metal parts used in assemblies can be used as-cast or as-struck. Many will require several subsequent machining operations before it can be put into service. Finishes and accuracy can be as different as a mold for an automotive cylinder head or a precision injection-molded medical part used to meter the flow of blood. But regardless of the part quality, better accuracy and finish of any die or mold will result in better replication of the part dimensions and longer service life of the mold or die.

VERTICAL EDM IN THE MOLD SHOP

Vertical EDM has been a mainstay in moldmaking operations for decades. It is also commonly called a "sinker" or "die sinker," derived from "hobbing." Hobbing is the process of forcing a preshaped, hardened die into a mold insert, cold-forming the resulting cavity. This was called "die sinking" and the term "sinker" EDM evolved naturally because now an *electrode* was sunk into the workpiece instead of a hardened die.

Moldmakers quickly learned the valuable time and material-saving advantages by using EDM for moldmaking. They could now EDM complex shapes into prehardened inserts while eliminating the costly procedures of inserting and laminating sections to get square corners or special shapes that conventional cutting tools could not produce. The accuracy and finishing capabilities provided by orbiter-equipped manual machines and CNC EDM further refined this craft. It is very rare today to find a toolroom engaged in full-time moldmaking operations without at least one EDM machine.

WIRE EDM IN THE MOLD SHOP

In the last few years, more and more wire EDM's are being installed in mold shops everyday. This has become one of the fastest growing market segments for wire EDM. Let's examine some of the reasons for this intense interest in this product.

ELECTRODE MAKING

Many years ago, this was one of the first uses for wire EDM in moldmaking. Early mold shops had vertical EDM machines with technology developed around copper electrodes. Since the early wire machines cut copper quite readily, this was a natural alliance. Complex copper electrodes could be easily and accurately wire-cut with no operator intervention. If multiple electrodes were required, accuracy from part to part was consistent and, unlike other methods, they never required deburring.

Early attempts at wire-cutting graphite electrodes proved to be very slow, but with the improvements in generator technology and the increased refinement quality of modern graphites, cutting speeds of up to 18 sq. in./hr. in premium graphite are now possible.

Rule: When choosing a graphite for wire-cutting, the best cutting speeds and finish will be obtained when cutting a premium, isotropic graphite with the finest grain and highest density possible. Often, the extra expense of selecting a better grade than necessary for EDMing is offset by the savings in fabrication costs wire EDM can provide.

Mold Base Work

Perhaps the second most common use of a wire EDM in a mold shop is to quickly and accurately cut the insert pockets in the holder plates of the mold base.

Question: Why reduce that large slug, where your insert is to go, into a bunch of oily chips at the expense of numerous end mills and machining time? Or...

Question: Why make a setup to drill four corners of a square or rectangular opening, then spend hours welding blades and band sawing the slug free of the mold base, then going back to the mill (again) for another setup?

Answer: Exactly my question... *why*? Why not wire-cut them?

Wire-cut insert openings are done precisely without the cutter whip and tapered openings so typical of milling operations. Square or radiused corners are easily produced using wire EDM, often enhancing the basic design and simplicity of manufacture while reducing total operations on the insert itself. Even if the original design didn't warrant through-hole pocketing, often the cost savings realized by cutting pockets in this manner can justify the expense of the extra back-up plate.

For those who make their own mold bases, if we can quickly and accurately wire-cut the pockets for inserts, why couldn't we also cut *all* of the other through-hole details in the mold base in the same setup? Other details, like holes for leader pins, return pins, taper-locks, etc., can be dimensioned from the same datum and all holes and openings are effectively "line bored" to tenth accuracy for perfect registration. Granted, secondary operations such as counterboring would have to be made for pin shoulders and guide bushings, but no subsequent operation would require elaborate setups or jeopardize the accuracy of previous operations, relying primarily on the use of piloted counterbores.

CAVITY WORK

Despite the fact that the wire must pass completely through the workpiece, many times cavities can be wire-cut. Flexibility and vision in the design stage will allow the use of wire EDM with substantial cost savings in machining and polishing time.

For example, a plastic tumbler or container mold, conventionally machined, might be designed in a simple two-plate mold configuration. This requires that, while soft, the insert be turned on a lathe or milled using tapered end mills (for draft), often requiring considerable polishing after heat treat.

This same job could be wire-cut "in the hard," and could be done almost totally unattended to better-than-required dimensions that will need little or no polishing. The outside diameter of the cavity (including draft) can be cut in the B-plate insert by allowing stock for the bottom radius of the tumbler in the A-plate insert (see Figs. 15.1 and 15.2).

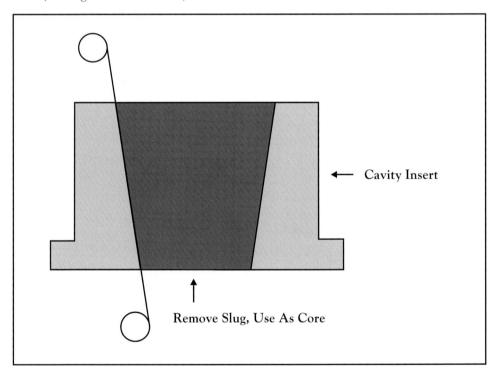

Figure 15.1   Creating a plastic tumbler mold, step one.

In the assembly shown in Figure 15.2, the radiused bottom of the tumbler can be turned or milled conventionally or vertical EDMed. Since this type of mold would most likely have a hot-runner system, the hardened, removable insert makes manufacturing and maintenance much easier and efficient.

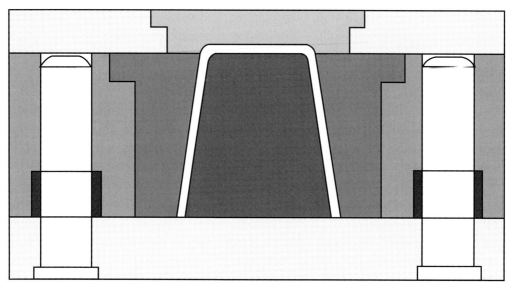

Figure 15.2   Final details for creating a plastic tumbler mold.

CORE PINS AND INSERTS

Many types of core pin shapes lend themselves to wire-cutting. Connector molds, for example, typically require cores with multiple, close-tolerance laminations that are quite labor-intensive and therefore expensive to grind and assemble. Many hours are required to dress numerous grinding wheels with the proper geometry to be used in sometimes elaborate setups to provide necessary draft angles. Better results can be achieved faster, in fewer setups, and with much less operator presence using wire EDM.

CORE PIN AND INJECTOR PIN HOLES

Many moldmakers use wire EDM to finish *all* through-hole work in cavity inserts. One method becoming popular is: after drilling all the necessary start holes, mounting holes and any rough cavity machining, the part is hardened. After the cavity itself has been finished, it can be completely polished without undue concern over minor bell-mouthing or slight rounding the edges of the ejector or core pin holes because they are not finished.

The final operation on the nearly finished insert will be the wire-cutting of all the ejector pin and core pin holes. These openings will be straight, to size, and will have dead-sharp edges at the cavity for minimal witness marks and no possibility of flash.

RELATED MOLD COMPONENTS

Remember the slugs that were left after wire cutting the insert pockets? Many times, especially if the material is of cavity steel, these can be used as the core, saving money in both raw materials and heat treating. If the slug material is from the mold base and can be hardened or heat treated, these can be used as structural

mold components such as slide bodies, locks, and wedges, etc. Any of these practices can cut costs by reducing or eliminating the purchase of additional raw material because the slugs can become the core or other components instead of oily
chips and wasted time.

SHUTOFFS AND STEPPED OR CONTOURED PARTING LINES

Besides the obvious applications of wire cutting stepped or contoured parting
lines on cavity inserts or mold bases, many times difficult shutoff details can be
wire cut on core pins, inserts, and slide components. Even long or delicate parts
can be successfully machined because, unlike conventional milling or grinding,
wire EDM imparts neither cutting pressure nor heat from friction to deflect the
part or to adversely affect the part's temper or surface integrity. Besides providing
flash-free shutoffs, these details can be executed time and time again with the same
consistent results, often times running almost totally unattended by an
operator.

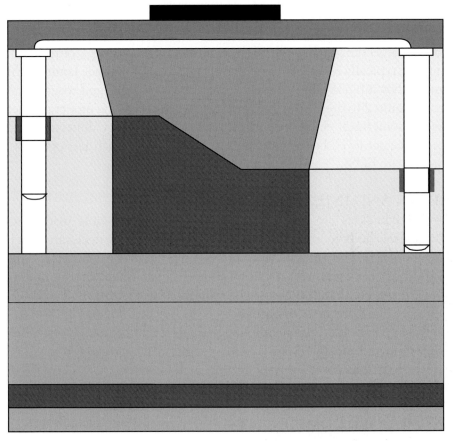

Figure 15.3   Molds with tapered inserts can be produced with wire EDM.

Another little-used potential realized when using wire EDM to produce molds is the ability to use tapered inserts and pockets (see Figure 15.3). Both can be wire-cut with ease and accuracy. Inserts tapered in this manner need no shoulder, heel, or mounting screws to secure them, but when assembled they are securely locked into place. Besides making mold assembly and maintenance faster and easier, it can greatly simplify design problems often encountered when considering optimum water placement or hot-runner configurations often complicated by mounting screw placement. This is another example of time and money saving potential that can be realized by using wire EDM in a mold shop.

REDUCE OR ELIMINATE POLISHING

Here's where it can all come together (or apart) — on the polishing bench. Unlike all the other operations it took to get the mold inserts this far, mold polishing is comprised almost entirely of labor-intensive *handwork*. True, the modern mold polisher has many mechanical devices to aid in this difficult and time-consuming craft, but from the most sophisticated ultrasonic polisher to the coarsest EDM stone, all of them have one thing in common — they all must be manipulated *by hand* — a hand that belongs to a highly skilled craftsmen. The mold polisher must be highly skilled because many an insert or detail has been scrapped on the polishing bench by an unskilled or inexperienced hand or eye. This skill is acquired in one way — through experience, and experience costs money and can only be charged against direct labor. To reduce direct labor costs and eliminate the possibilities of incorrectly polished or damaged details, wire EDM can save enormous amounts of time and money while relieving the pressure on this most common bottleneck in mold production.

Figure 15.4   Mirror finishing capabilities augment the moldmaking industry. (Courtesy of Sodick, Inc.)

The typical wire EDM finish is often satisfactory and can be run "as is," after only two or three skim cuts on a standard power supply, while some machines provide fine-finish options (<0.5 μmR$_{max}$) that provide for true mirror finishes. Should any further processing be necessary, such as draw polishing or optical-quality finishes, the recast or "white-layer" left by wire-cut EDM is minimal and lends itself quite easily to further refinement with a minimum of time and equipment.

SUMMARY

Hopefully, in the span of this chapter, a nonuser can come away with a greater understanding of the potential savings derived by exploiting the capabilities of a wire EDM in a mold shop. In this chapter's brevity I am certain that some salient points have been missed. But, to review, here are nine of the important advantages of having a wire-cut EDM in your mold shop:

1) Electrode making
2) Mold base work
3) Cavity work
4) Core pins and inserts
5) Core pin, ejector pin and ejector sleeve holes
6) Related components
7) Shutoffs and stepped or contoured parting lines
8) Reduce/eliminate polishing
9) Significantly reduce direct labor costs.

These examples alone should be enough for you to consider examining how a wire-cut EDM can benefit your moldmaking operation.

Just as is done with vertical EDM, all aspects of mold manufacture should be examined to determine if part of the solution could be wire EDM — or wire EDM in conjunction with vertical EDM. This has already proven to be a very good marriage. While vertical EDM in moldmaking is engaged primarily in cavity work, wire EDM is applicable in almost all aspects of moldmaking: cores, cavities, electrodes, mold base, work and components. The combination of the wire EDM's flexibility and ability to run unattended make it an ideal addition to any moldmaking facility. Of this, there can be no question as purchases of wire EDM's by mold shops makes it the fastest growing segment of manufacturing today.

Looking back over the list of some of the different moldmaking disciplines at the beginning of this chapter reminds us that moldmaking of *any* type is both an art and science and therefore, will never be easy — but it can be made *easier*. Learn the advantages that a wire EDM can bring to your shop. Discover how you can wire-cut difficult and labor-intensive parts and details unattended to split-tenth accuracies while often bypassing the polishing bench entirely.

Sound easy? Well, it's at least eas_ier_.

Chapter 16

PREVENTIVE MAINTENANCE

INTRODUCTION

Wire-cut EDM sales in America continue to rise, which is good for the builders of the equipment and for American manufacturing as well. When sales are good, the builders can continue to confidently invest more of their resources into research and development as they try to further improve upon EDM technology. The users of this technology, from job shops to Fortune 500's, use this technology to machine their parts faster, hold tighter tolerances, improve finishes, reduce part rejections, etc. This is a very efficient and self-perpetuating symbiosis of builder and user, of manufacturer and manufacturing.

Because of this technology, today's wire-cut EDM's can be incredibly productive and profitable machine tools. With this modern technology they can not only routinely do what was previously considered impossible, but they do this with increased autonomy and efficiency. More and more wire machines are running unattended far into the night or over weekends, further increasing throughput and profitability of their users.

However, all this capability can only come with a price. It is a small price that is above and beyond the cost of the machine itself, but one required without exception. This price is maintenance and preventive maintenance. Because these machines have the ability to run long hours unattended, we can easily take for granted just how many hours the machine has been running. For this reason, it is highly advisable to make and follow a written timetable or schedule of maintenance procedures.

There are six main systems or component groups of a wire machine. They are as follows:

1) The machine tool.
2) The wire-feed mechanism.
3) The tapering device.
4) The automatic wire threader.
5) The power supply and control.
6) The dielectric system.

We will examine each group of components that make up a wire EDM machine, and make recommendations for their care. Considering the different makes and models and ages of equipment already in the field, the following procedures and recommendations can only be used as guidelines. Therefore, these instructions will be kept very generic for obvious reasons, although paying particular attention to component cleanliness is very important and is universal to all makes.

For the very best results or for instructions specific to your machine, consult the manufacturer's machine manual whenever possible.

THE MACHINE TOOL

LUBRICATION

Follow the specific recommendations found in your machine's maintenance manual. Some machines are equipped with manual oilers, while others are electric and therefore their operation is automatic. Obviously, both types must be kept full of whatever lubricant the builder has specified.

Some machines do not use lubricating oils but must be greased either by manually packing the slide ways and screws by hand, or by forcing grease into working mechanisms through zerk-type fittings. Refer to your user's manual to learn where these are.

Some of these machines will require the removal of one or more sheet-metal panels and skirts to gain access to some of the slide ways and grease fittings. Since this type of service is required only once or twice a year, *remove all of the panels necessary* to gain access. It may seem like a nuisance, but if you don't do this, a serviceman will eventually have to remove these very same panels to replace the worn components that failed from lack of lubrication.

(*Note:* With so many makes and models of equipment in place, it would be impossible to specify the specific types of lubricants and their intervals in a generic manner. Make sure that this procedure is done exactly as the service manual recommends.)

THE WIRE-FEED MECHANISM

While the first two topics of this section do not exactly fall into the realm of preventive maintenance, I don't believe readers will mind a slight diversion concerning job preparedness. This concerns both the supply of fresh wire to the machine and the take-up or disposal of the spent wire.

WIRE SPOOL

When planning long-running or overnight jobs, make sure that the wire spool has enough wire to complete the job. This may sound elementary, but on occasion operators have planned for a job to run unattended or overnight but failed to provide enough wire to do so. The machine will run out of wire before the job is completed, and this will delay the completion, delivery, and billing of the parts.

Entry-level operators are advised to load a fresh spool for overnight operations until they are familiar with calculating the time remaining on a spool of wire. Calculate the amount of wire remaining on a partially used spool by its diameter, weight, and the approximate wire speed. Figure 16.1 should help give an accurate estimate of wire time remaining.

With more experience, a mere visual inspection can provide a reasonable estimate of the wire remaining.

WEIGHT x SPOOL SIZE = TOTAL WIRE LENGTH x WIRE SPEED = TIME.

DIA.	=	Yds. per lb.	×	3.5	6.6	7.7	11.0	13.5	35.0	44.0
0.006"		3,300		11,550	21,780	25,410	36,300	44,550	N/A	N/A
0.008"		1,864		6,524	12,302	14,353	20,504	25,164	65,240	82,016
0.010"		1,193		4,185	7,874	9,186	13,123	16,105	41,755	52,492
0.012"		820		2,870	5,412	6,314	9,020	11,070	28,700	36,080

Figure 16.1   Yards of wire by spool weight (approx.). Data for constructing wire chart courtesy of Gisco® Equipment, Inc.

WIRE BASKET/TAKE-UP SPOOL

This is actually the second half of providing sufficient wire to complete the job — do not let the wire basket get too full. This can wind the spent wire into snarls or "bird's nests" in the pinch rollers or belts. Besides stopping the machine, this can create a costly repair bill.

Likewise, on machines equipped with a spent-wire take-up spool, make sure it doesn't become wound with too much wire. On some machines, too much wire on the take-up spool can actually cause jamming of the spool within its recess, stopping the machine. On reoccurring or production jobs with a known wire consumption, this can be calculated into a maximum safe time between wire removal. Although having to return during the night or weekend to remove spent wire from a take-up spool is annoying, it's better than having a jammed spool and a possibly delayed delivery of the job.

PINCH ROLLERS, BOBBINS AND BELTS

All of the rollers, bushings, bobbins, and tensioning devices in the drive train can fall into this category. They should all be cleaned on a regular schedule and even more often when coated or stratified wires are used. Check for wire grooves that will eventually wear into all rollers, belts, drive, and idler surfaces.

Check all wire-drive belts regularly for wear, cracking or fraying. Always keep spare belts in stock and change them at the intervals prescribed by the manufacturer.

POWER-FEED CONTACTS/ENERGIZER PINS

Inspect these often, sometimes daily, depending upon their use. High-speed (high-power) machining will wear the contacts much faster than skim cutting will. Gradual loss of cutting speed, erratic cutting behavior, increased wire breakage, or "brassing" or "plating" of the workpiece are all signs that the contacts could need attention. Index or rotate them at regular intervals (or sooner), depending upon the nature of the cutting to obtain maximum speed and efficiency.

WIRE GUIDES

Inspect weekly, or more often when using coated wires or cutting tapered parts. Check them with a loupe or microscope. Remove brass deposits or zinc buildup by immersing the guide assembly in an inexpensive ultrasonic cleaner. On machines using split guides, make sure that all mating surfaces of the guides themselves and the actuating mechanisms are clean and free of wire shavings or debris.

FLUSH CUPS/NOZZLES

Inspect daily. Check for cracks, fractures, or any kind of damage, particularly around the flat, "sealing" surfaces. These areas are critical to efficient flushing and cutting speeds. A cracked or chipped nozzle can reduce high-speed cutting speeds over 30% and can substantially increase the probability of wire-breakage. "Flushing, flushing, and flushing" is just as critical to wire EDM as it is to vertical EDM.

OPTICS

Some machines utilize an "optical-guidance" system to enhance their cutting ability. Follow the builder's cleaning and maintenance instructions carefully to ensure that you receive all of the benefits and performance enhancements you paid for.

TAPERING DEVICES

U-V DRIVES

This is the device used for generating tapered cuts and it is named after the U and V axes, respectively. There are several different types, depending upon the age and design of the machine.

Some designs use lead screws to move the upper control arm around. These are preloaded against backlash by friction using split-nut or collet-type devices. Since friction causes wear, adjustments must be made to these types more frequently than ball-screw types. Follow the recommendations of the builder for lubrication and maintenance intervals.

Most types contain the servo motors, drive screws, and nuts within a box located on the upper control arm. Sometimes these units can leak and fill with water, quickly rusting the components and severely affecting tapering accuracy. These units must be entirely rebuilt and thoroughly sealed with an RTV sealant. The use of a sealant should reduce the need for complete rebuilds or replacement.

Newer designs by several builders have placed the U-V mechanisms up and away from the proximity of the water, reducing the need for water-related precautions and maintenance.

Some machines use articulated heads instead of a U-V box. These heads contain the wire guides and they must gimbal or swivel on their mounts to produce tapers, so free movement of these components is very important. Make sure that all linkages and related bearings are clean and well lubricated.

Another machine design eliminates the U-V device entirely. Tapering is accomplished by using a duplicate set of control arms similar to the X-Y. This allows the wire to travel instead of the worktable. With this design, maintain the tapering axes (U,V) just as you would the positioning (X,Y) axes.

AUTOMATIC WIRE THREADERS

GUIDE PIPES

The high-pressure jet of water during rethreading is meant to center the wire in the pipe for a more accurate try at the start hole below. These need to be clean and free from any bends or kinks. The flow of water leaving the upper nozzle or guide pipe can be visually checked for a smooth, uniform flow that is centered in the opening. Any water spray, visible turbulence, or deflection of flow from center is indicative of a problem.

PARTING MECHANISM

When moving from part to part in production operations or detail to detail in die work, the wire machine must sever the wire to enable it to pass on to the next threading position. There are two types of parting mechanisms — cutting and annealing.

The cutting type uses sliding dies made of carbide or ceramic to mechanically cut the wire. One type holds the wire captive within a hole while an actuator slides a square-edged cutter across the top of it. Other types use sliding cutters (some round discs, some square lozenges) in the same manner. The principle is the same for all of these, and the mechanisms must be kept clean and the cutting edges must be very sharp. Clearance between the shear and anvil must be very close or burring or flattening of the end of the wire will result. This can contribute to wire misfeeds, reducing threading reliability.

Annealing-type mechanisms work by applying an electrical current through a small section of pretensioned wire. As the wire is heated to the point of melting, pretension or another mechanism pulls the heated section apart, stretching it to a perfect pinpoint for ease of rethreading. If you are familiar with the way a band saw blade welder operates, essentially the *opposite* procedure occurs in the annealing mechanism of a wire threader.

All electrical contacts and any mechanical grippers of this device should be cleaned regularly and even more often when using coated wires.

AIR REGULATOR

Most automatic wire threaders use compressed air. Check the pressure gauge of the air regulator monthly. Check and drain the condensation reservoir weekly, and more often in humid areas where large amounts of water tend to accumulate in the air lines.

THE POWER SUPPLY AND EDM CONTROL

MAIN POWER CABLES

Inspect the power cables monthly. All cable connections should be free from rust and corrosion. Check for clean, secure connections at the machine tool and at the power supply terminals. Avoid bending, pinching, or crushing them between moving machine parts or workpieces. Power cables should always be free from any breaks or kinks. If a connecting end of a power cable becomes frayed or broken, replace the entire cable. *Never* attempt to lengthen or shorten a power cable. Their length has been carefully engineered and precisely controlled to obtain the correct electrical impedance required by the power supply. Any modification of any kind to a main power cable can significantly change all known cutting characteristics and machine performance.

POWER SUPPLY CABINET

Visually inspect all sealing surfaces of cabinet panels upon their removal to insure an air-tight seal. There should be an imprint pressed into the rubber seals indicating a good seal when the mounting screws are tightened.

The intake fans used for cooling the cabinet are powerful and they should draw air from only the filtered intake screens. To ensure adequate air circulation within the cabinet, make sure that exhaust ports are clean and free from obstructions. Some machines exhaust their air circulation through louvers or vents in the top of the cabinet. Do not inadvertently block these openings by placing blueprints, clipboards, or other objects on the cabinet top.

INTAKE SCREENS AND FILTERS

Inspect and clean all intake filters and screens periodically. This interval should be listed by the builder within their manuals, but this can also be determined by operators and maintenance personnel depending upon shop conditions.

More frequent attention would be warranted if the shop environment involves other machine tools that are utilizing coolants and/or cutting fluids that can result in their vapors being drawn into these intakes. Also, grinding operations can suspend microscopic metallic particles in the air and these are especially hazardous to electronic circuitry. If these conditions cannot be prevented or controlled, frequent attention to this area is highly recommended.

CIRCUIT BOARDS

Most shop environments exercise reasonable control over airborne vapors and particles. Even so, sometimes machine panels don't seal entirely, or submicron particles of airborne dust pass through intake filters. In this case, a yearly check of printed circuit boards (PCB's) is sufficient. In areas where coolant mists and/or grinding dust are prevalent, more frequent inspection and cleaning are required.

If necessary, carefully remove the dirty circuit boards *one at a time* to avoid any mix-ups and gently clean any dust or black "electronic fuzz" from all components and circuitry with a clean, soft-bristle brush. An inexpensive, two-inch-wide paint brush works fine for this.

Another method that works even better and keeps the shop clean at the same time is to use a vacuum cleaner with the round, soft-bristle brush attachment used for furniture. Do any cleaning *gently* to prevent bending, damaging, or dislodging any component on the board itself.

Caution!! <u>*Never*</u> *use an air hose on circuit boards or anywhere within the cabinet!!*

Before reinstalling the newly cleaned board back into its slot, gently clean the contact tab(s) with a soft cloth and a contact cleaner designed specifically for this purpose. *Do not* rub the contacts of *any* PCB with a pencil eraser, Scotch-brite, or any other kind of abrasive media. The abrasives can rub through or damage this fine plating and actually defeat the purpose of achieving good contact.

DISK READERS/DRIVES

To prevent data loss or distortion, it is recommended that each reader or drive be cleaned with a disk cleaner recommended or supplied by the builder. Commercially available cleaners should work fine also, but check for compatibility just to be sure. Cleaning should be done at least once each month, and more often if frequently loading or downloading programs or running directly from disks.

SOFTWARE

Make frequent backups of program disks. Besides the occasional, rare disk failure and the resulting loss of information, fluctuations of incoming power while editing or saving data can scramble or erase valuable programs. If this is a common occurrence (more than twice a year), and it is not attributed to disk failure, it is strongly suggested that obtaining voltage suppression may be warranted. (This is protection that transformers, voltage regulators, or "a dedicated line" cannot provide.)

Duplicate *all* disks and store them in a secure area. This will safeguard against job-program loss as well as actual EDM system data loss and perhaps the resulting inability to use the machine. EDM system-data can consist of cutting conditions, location and assist files, offsets, screw and pitch error compensations, etc. This type of information is integral to the operation of the machine, and you should always have backup data available. Without a backup disk, one would not only need the latest printout of this information, but all of these data would have to be manually reentered into memory. This amount of data entry would be very tedious, time-consuming, and it carries a high probability of error.

Three copies of the machine's system data or "brains" should be made: one for the machine's daily use, an operator's backup copy stored in a compartment of the machine or a nearby workbench drawer, and a third copy to be stored in a secure area away from the machine in an office or where the machine's records are stored. Multiple backups provide good insurance against the loss or damage of the originals.

THE DIELECTRIC SYSTEM

DIELECTRIC RESERVOIR

This is simply the tank or sump for the storage of all water used by the wire machine. It requires almost no maintenance except for periodical checks for rust and pin-hole leaks especially in older units.

Actual machine use in combination with recommendations from the builder should determine how often one should entirely replace the water. When changing out the water, drain the tank completely and flush all chips and sediment from the bottom. Wipe all residue of carbon "soot" from inside the tank walls and rinse thoroughly.

While doing this, you may notice that the tank walls may also have a slippery feel to them. This is caused by bacteria growing in the water. This bacteria can cause premature resin or filter failure or cause the water to emit a stale or musty odor. By completely removing this film, new bacteria will have a smaller culture or colony to grow from, so any new growth will be inhibited.

To help control this condition, special biocides can be added to the fresh water, but they may also affect conductivity. Special ozone generators have been developed for EDM applications. These devices create conditions that oxidize organic matter within the dielectric system while preventing future buildup of stale-smelling or slimy water.

DIELECTRIC LEVEL

Check the water level frequently, especially when making high-speed cuts and using high-pressure flushing which can consume large quantities of water. Most tanks or reservoirs have a float-level indicator or a clear viewing tube along its side. Do not allow your pumps to run dry. Besides the nuisance of having to stop to prime them, it can also shorten their expected life.

(*Note*: Some wire machines use a dielectric oil instead of water. Still others use a combination oil/water dielectric system The oil-only machines will not have the evaporation problems typically found with water machines, but careful attention must be paid to separators, filtration and resin systems of machines with oil/water capability. Since water-only machines dominate the market, we will address only this type. Specific information regarding oil-only or oil/water dielectric systems is available directly from the builder of these machines.)

FILTERS

Paper-media filters are in use on almost all major wire EDM's. They are used to separate the chips and debris from the water that have been generated by the EDM cut. To maintain the best speeds, finishes, and accuracies, regular filter changes are mandatory. If you are cutting the same type of material daily or running production, you can easily determine filter life and schedule changes accordingly. Otherwise, watch the water-pressure gauges carefully and change filters as soon as it is needed. Many controls have maintenance screens to assist in this scheduling. Dirty or clogged filters can leak or blow out, contaminating the entire dielectric system with a high probability of shortening the life of your resin (see "Resin," below).

Some wire users have installed prefilters on their machines to extend the life of their machine filters. Follow the recommendations of the manufacturer of this system for proper maintenance. Likewise, if you have equipped your machine with a central filtration system, consult the manufacturer for specific information on its required maintenance and intervals. Central-filtration systems are typically very reliable, but you must follow their instructions explicitly because if the system fails, not just one machine will be down, *all* of your machines will be down — not a pleasant thought.

Whether your filtration system requires replaceable filter canisters, diatomaceous earth, or other media, it is strongly advised to always have everything necessary for one complete filter change in stock at all times.

RESIN

Mixed-bed resins are used to control the chemical cleanliness of the dielectric. Careful monitoring and changing of the filters will prolong the life of the resin. The resin itself is in the form of small plastic beads that have approximately 65% of their ion exchange surfaces on the inside. Unfiltered water can clog these very small capillary passages and significantly shorten the useful life of the resin.

Conductivity meters or resistance sensors on your machine will alert you to the resin's condition. Like filters, a reserve supply of resin should be kept in stock for timely replacement.

CONDUCTIVITY/RESISTIVITY SENSOR

This is a small sensing device with two small electrodes immersed in the deionized water. A low current is passed across these probes to measure and indicate the water's conductivity (or resistivity). Depending upon the water quality, mineral deposits or corrosion can influence sensor readings. Check and clean the sensor probe(s) periodically.

HOSES

These require only periodic checks and are all typically long-lived, as long as good water quality is maintained. High-pressure hoses are usually clad with stainless-steel braid and almost never fail. Never allow them to become bent or kinked. Replace any braided hose that has signs of fraying, especially at a fitting.

Rubber or silicone hoses should be checked monthly, both visually and by squeezing them with your hand. Visually check for seepage or leaks at all fittings and junctures. By physically squeezing a hose, you should be able to determine if it is hardening or softening with age. Neither condition is desirable, and should this occur, they should be replaced.

VALVES AND SOLENOIDS

Valves and solenoids seldom require attention if good water quality is maintained. Mechanical gate valves and bibs should be periodically opened and closed throughout their entire range to ensure that gradual buildup of mineral deposits will not jam them after prolonged use.

This procedure should be done with the system pumps off to prevent the possible rise in water pressure from damaging filters, hoses, or other components. Also, before adjusting any control valve, note its position or gauge setting before starting so it can be correctly reset when finished.

Rust, salts, and corrosion can cause electric solenoids to stick — sometimes open, sometimes closed. A tap with a soft-faced hammer is usually enough to dislodge the chip or debris that is blocking its movement. While this method is not very scientific, it usually works if the blockage is not severe. If this sticking problem is reoccurring, the errant valve may require disassembly and cleaning. Most solenoids are simple devices, and cleaning usually solves this problem, although some are not meant to be repaired and must be replaced.

Important!! Before you start "repairing" solenoids or any other machine component, check with the machine builder to make sure that you are not endangering your warranty by working on it yourself.

GENERAL

Preventive maintenance doesn't stop with the machine tool. There are other things you can do to get the most from your EDM equipment. Some of these are the following:

DOCUMENTATION

All records, parts receipts, service reports, internal memos, etc., pertaining to the machine's performance or maintenance should be copied several times and stored: 1) In or near the machine itself, 2) In the Maintenance Department file (if applicable), and, 3) Wherever the original purchase order agreement is filed.

Redundancy in this area is to ensure against their loss or destruction. This is to document the machine's complete history from the time of installation. This can provide important data that can assist with service or troubleshooting by revealing possible trends in machine performance. Documenting the machine's history can prove useful for the possible trade-in or resale of the machine in the future.

Documentation also includes the builder-supplied machine manuals. If you are lacking any book, manual, or instructions for a machine, call your distributor or machine manufacturer. If your machine is older or is no longer being made, contact someone in your area with the same equipment to see if they will give or sell you copies of their manuals.

SERVICE AND SUPPORT

If at any time there is a question or problem pertaining to any aspect of your machine's operation, performance, or maintenance, do not hesitate to contact your local distributor or machine manufacturer. Keep business cards with the telephone and fax numbers of local service and applications engineers near the machine for easy operator access.

GENERAL CLEANLINESS

The three words, "keep it clean," cannot be emphasized enough. We already know the logic and value of keeping the mechanical components clean, but this applies to the sheet metal cabinets and machine castings also.

Besides all of the known advantages that a clean and properly maintained machine provides, a larger, psychological condition exists when a machine and its surroundings are clean. Workers tend to perform better, more efficiently. They will generally take greater pride in their accomplishments and have better attitudes toward their craft. All of these gains will show up with higher productivity, better quality, and safer operations.

SUMMARY

Early on in this chapter, we established that an additional price must be paid for machine performance that is above and beyond the actual cost of the machine. Obviously, preventive maintenance does carry a price tag — the price of the consumable items or replacement parts themselves and of course the cost of the time it takes to perform these tasks. While this does require an outgo of capital, real or potential, you must not consider this an expense.

Preventive maintenance is the best insurance toward protecting your investment.

Don't be penny-wise and dollar-foolish. Don't try to stretch the time between known maintenance periods to save a few pennies now and wind up paying many dollars later for expensive repairs that could have been avoided. Schedule the time and money to perform these simple tasks before they become much more complicated.

In many respects, a wire EDM is like a high-performance race car. Although it can be equipped with numerous options to improve its performance and beat the competition, the best and fastest car on race day *will not even finish* without proper maintenance. Neither will your wire EDM. Auto racers wouldn't think of skipping or scrimping on maintenance and neither should you.

The key word in preventive maintenance is *preventive*. By performing regular *preventive* maintenance, we can significantly reduce or eliminate most *unnecessary* maintenance. If I had to further define or describe preventive maintenance in a single sentence, I would recite a jingle from an old commercial (coincidentally an automotive product):

"You can pay me *now*, or pay me *later*."

SECTION IV

ADDITIONAL INFORMATION

Chapter 17

THE EDM FINISH AND SURFACE INTEGRITY

INTRODUCTION

One of the most misunderstood aspects of an often misunderstood manufacturing process is the EDM finish. What is it? How is it different from "conventional" finishes? How do you measure it? What is recast? What is its effect on surface integrity?

These are only a few of the many questions raised when manufacturing and process engineers consider using the EDM process. Just like most other facets of EDM, numerous variables can affect the desired results — some favorably, some not. While trying hard not to conduct a class on metallurgy, we will attempt to arrive at a better understanding of the EDM finish itself and its relationship to material surface integrity.

EDM THEORY AND EDM FINISH

With few exceptions, most of us don't care for reruns on television, so I won't push my luck by reciting "EDM 101" again. But to know and understand the EDM *finish*, you must know and understand EDM *theory*.

In the briefest of summaries, each individual spark vaporizes, then melts, a small amount of workpiece material and leaves a crater in the workpiece. This is consistent whether it is wire-cut or vertical EDM. High power and long spark duration (low frequencies) provide high metal removal rates and rough surfaces. Obversely, the combination of low power and short spark duration (high frequencies) results in finer finishes. Rough or smooth, we encounter our first difficulty — *measuring* this finish.

MEASURING THE FINISH

There are several ways one can measure the finish of an EDMed part. Almost all builders of EDM equipment can supply a metal or plastic standard with different finishes for visual inspection. Others use the age-old "fingernail profilometer" — dragging a fingernail across the test piece and comparing that "feel" with the actual part. Neither are precise enough to accurately "measure" the surface finish of a critical part.

A profilometer (or surface roughness gauge) is a device we use to measure the surface of a machined or polished part. It is a very precise electronic instrument that utilizes a diamond stylus that is traversed across the surface, tracing it. The movements of the stylus caused by surface roughness, waviness, and irregularities are, after filtering and damping, converted into electrical signals by the transducer, and number values representing finish are displayed on an LED or LCD readout.

A newer and less common type of profilometer is of the noncontact type using a laser beam. This device would be used on surfaces that could be damaged by the diamond stylus or prove damaging to the stylus itself. Here's how it works. A sharply focused beam of laser light, about 0.1 micron in diameter, is directed onto the workpiece. Surface finish is calculated by measuring the amount of movement made by the measuring arm while it maintains the correct focus and diameter of the laser beam.

Another method of measuring finishes is by using a combination of visual and physical measurements via screen images or photomicrographs from a scanning electron microscope (SEM). The finish can be measured optically by viewing the part through a graduated measuring grid or template, or by physically moving the part under a graduated or cross-hair reticle on a very accurate, graduated, screw driven table or stage similar to that of an optical comparator.

FINISH MEASURING SCALES

First, what *scale* do we use to measure an EDM finish? Unfortunately, the engineering community is still quite divided on what scale to measure finish by. Most of us are familiar with surface measurement by RMS (Root Mean Square), but "technically" RMS has been obsolete for some time although it is still seen on many blueprints today. Thrown in for confusion's sake are AA, R_a, R_{max}, μR_{max}, VDI, etc. If this seems confusing to engineers, pity the poor layman!

Figure 17.1 shows the definitions of some of the more commonly used finishes. They are measured in units of μin. (microinch) or μm (micrometer).

The United States standard for measuring finishes is R_a, which is an *arithmetic average* of the highs and lows of a workpiece surface. The highest peaks and lowest valleys of a surface are represented by R_{max}. Determining RMS (Root Mean Square) is a bit more complicated. It is obtained by first squaring the high and the low measurements and then averaging them. The square root of this average will be the measure of finish. A general rule of comparison is that RMS values will be about 11% higher than those measured by an arithmetic average.

SCALE	DEFINITION	UNIT*
RMS	Root Mean Square	μin.
AA	Arithmetic Mean (average)	μin. or μm
R_a	Roughness Average	μin. or μm
H_{max}	Maximum Roughness Depth	μin. or μm
R_{max}	Maximum Roughness Depth	μin. or μm

* 1 μin. = 1/1,000,000 inch or 0.025 μm
   1 μm = 1/1,000,000 meter or 40 μin.

Figure 17.1   Commonly used finish measurement scales.

Until the international engineering community and the National Institute of Standards and Technology can agree upon a single, *universal* finishing standard, we will be forced to use conversion charts like the one shown in Figure 17.2.

R_a μin.	RMS	R_a μm	VDI
0.5	0.55	0.012	—
1	1.11	0.025	—
2	2.22	0.050	—
4	4.44	0.100	—
8	8.80	0.200	6
16	17.76	0.400	12
32	35.52	0.800	18

Figure 17.2   Conversion chart for finish measurement scales.

COMMON DEFINITIONS OF FINISH PARAMETERS

R_a = **Roughness Average**. The average roughness R_a is the arithmetic average of all departures of the roughness profile from the center line of the evaluation length. It is also known as the arithmetic average (AA) and the centerline average (CLA).

R_{max} = **Roughness Depth**. The mean roughness depth R_{max} is the largest of five roughness depths (peak to valley height) from five successive sample lengths of the roughness profile.

RMS = Root Square Mean. The "perfect surface" that would be formed if all the roughness peaks were cut off and used in filling the valleys below this surface.

THE EDM FINISH

Even after deciding upon an acceptable finishing scale or standard, we still have some confusion concerning *measuring* an EDM finish. Part of this confusion is caused by the *lay* of the surface finish. *Lay* is technically defined as the "direction of a predominate surface pattern," or, in plain shop-talk, tool marks. Conventional "chip cutters" leave their "signature" on the workpiece in the form of distinctive tool marks consisting of microscopic, directional grooves, whether the material has been milled, turned, or ground.

We are all familiar with the markings of concentric circles left by a fly cutter or end mill; the smooth, straight "brushed look" that is produced by a surface grinder; or the radial patterns in material that has been blanchard-ground. The witness marks or grooves left by these cutting devices are *directional* and determine its lay. EDM does not *physically* cut chips, but removes materials *electrically* by random spark erosion. The surface finish produced by EDM is in the form of small, hemispherical craters with no distinct pattern or lay to influence measurement.

Measurement of a surface finish is typically made by tracing across the lay, or across its worst direction; for example, the tracing stroke of a profilometer would be made *tangent to* or *across* a surface-ground finish. Measuring *with* the lay of the grind would produce inaccurate readings. Since EDM finish has no lay, it can be measured from any direction or angle.

At this point, with seemingly everything concerning EDM finish being so different when compared to conventional finishes, it becomes obvious that we must examine not only the surface finish of a material, but also just as importantly, *the relationship of surface finish to the actual integrity of the EDMed surface*.

RECAST

After we have agreed on the finish desired and the scale it is to be measured against, we must contend with another aspect of the EDM process. This is something called the recast layer or "white layer." In EDM, regardless of finish, the smoothest, shiniest surface finish doesn't mean a thing if, during roughing operations, the surface integrity of the material below the finished surface has been compromised.

Here is a brief description of recast. The initial strike of the EDM spark will vaporize a small amount of material. After the initial strike and vaporization, any additional material will be melted due to the increasing surface area of the expanding crater and the sinking of heat into this now-larger surface area. Some of the molten material will be ejected or float away from the crater in the form of liquid globules that freeze into spherical "chips" upon contact with the much cooler dielectric. There will always be a portion of molten material that will not have sufficient mass to escape the effects of surface tension, and these will be drawn back into the spark crater during off-time to solidify or "refreeze" back onto the cooler crater walls. This resolidified material is called "recast" and, depending upon the dielectric (water or oil), this layer, metallurgically, can be totally different than the parent material.

OIL AND WATER

There is a distinct difference between the recast layer left by wire EDM and the recast left by vertical EDM. This is attributed mainly to the different dielectric fluids used: special oils for vertical machines, and deionized water for wire-cut. Likewise, the effects upon surface integrity are as different as oil and water.

Oil dielectrics can change the parent material by creating an unintentional and wildly uncontrolled heat-treating process: super-heating the area, then (during off-time) quenching it in oil. The high heat of the spark breaks down the oil into hydrocarbons, tars, and resins. While in a liquid state, the molten metal draws the carbon atoms from the oil and they become trapped within the recast layer, creating a "carburized" or martensitic surface. Either term correctly describes the recast surface as having a high carbon content. While these results are a far cry from the carefully produced carburized surfaces obtained intentionally during professional heat-treating operations, it is nevertheless hard — glass-hard, in fact. Even with materials that have been prehardened, the recast layer produced in oil can be several Rockwell points higher (C-scale) in hardness than the parent material. Ask anyone who has stone polished an EDMed cavity how hard the recast layer can be.

Finished surfaces produced in a water dielectric can be several points lower in hardness than the parent material. This is because oxides are produced by the vaporizing water. This oxidation, along with electrolysis, can release or deplete carbon, cobalt, and other atoms from the material's surface. This can also make skim cuts more difficult in aluminum, molybdenum, and other materials because the reduced power levels used in skimming cannot always overcome the increased resistance of the oxidized surface. In addition, copper atoms that have been released from the brass wire can be assimilated into the briefly molten workpiece material, becoming part of the recast layer, contributing to an uncontrolled alloying process, further influencing the parent material.

As described above, the recast layer in oil tends to be highly carbonaceous, depending upon the material's affinity for carbon (carbon atoms are released from the oil by the high temperature of the spark and can be absorbed by the molten metal). In water, the heat from the spark "cracks" the water and breaks it down into its principle elements — hydrogen and oxygen. The exterior surface left by EDM and a water dielectric is typically an oxide of the parent material, produced by the oxygen-rich area around the spark. Certain materials with an affinity for hydrogen, tend to absorb hydrogen atoms and undergo a phenomenon called "hydrogen embrittlement," sometimes jeopardizing its structural and/or surface integrity.

Regardless of the dielectric used or the surface's actual metallurgical composition, this thin strata of resolidified material is called recast. It is also referred to as the "white layer," so named due to its whitish appearance in photomicrographs. The recast layer is an inherent byproduct of the EDM process and it is unavoidable. Although today's sophisticated new power supplies and improved generator technology now provide a very fine degree of control resulting in a significant reduction of topical metallurgical damage, the thermal nature of the EDM process itself makes it impossible to eliminate recast entirely.

Just below the white layer is the area called the "Heat Affected Zone" (or HAZ). This area has been only *partially* affected by the high temperatures of the EDM spark. The thickness layer of the recast zone and the HAZ immediately below it depends on the current and frequencies used during machining, and the ability of the material to conduct and transfer heat away from the machined area. Depending upon the material and the temperatures reached, the HAZ can be several points lower in hardness than the unaffected parent material. This is due to the "drawing" or tempering effect, caused by elevated temperatures. Both the recast layer and the HAZ can affect the structural and/or surface integrity of the EDMed area.

In cases of severe surface influence or structurally sensitive parts, recast removal and/or stress relieving of the part is sometimes necessary. If the recast layer is too thick and is not reduced or removed by some manual or mechanical polishing — Abrasive Flow Machining (AFM) or Electrical Chemical Machining (ECM), etc. — this often very hard and brittle recast layer can cause cracking or flaking of the machined surface, impart stress and stress risers, and contribute to premature failure of the part. For years, this has rightfully been a major concern of the aerospace and aircraft industries; however, with the continued improvements and refinements in modern EDM power supplies, the acceptance of EDMed parts for aircraft and aerospace industries is becoming more and more routine.

Although the EDM process is realizing greater acceptance and success in the aerospace and aircraft industries, there is still a pressing need for better education of the engineering community to overcome the traditional resistance this process has encountered. This is in no way to be interpreted that we in EDM are casting disparaging comments about the reluctance of aerospace and aircraft manufacturers to embrace our craft — quite the contrary. Since EDM is *our* discipline, *our* craft, and *our* forté, it is logically *our responsibility* to educate and demonstrate to them that ours is not only a viable and cost-effective method of manufacturing, but that it is also safe, manageable, and predictable.

CARBIDES, EDM, AND ELECTROLYSIS

Carbides can be EDM machined, but improper power settings can damage the workpiece by depleting the cobalt binder material that holds the tungsten particles in place. This can result in flaking or cracking of the EDMed surface. Fortunately, most modern power supplies are capable of reducing most of this concern of controlling the thickness of the recast material by constantly monitoring and controlling the spark/servo conditions during high-frequency machining. One specific reason for the improvements in the surface integrity of EDMed carbides, especially in newer wire-cut machines, is the reduction or total absence of capacitors. Pure transistor-controlled machining, while perhaps not cutting as fast, is much more precisely controlled and is far less damaging to carbides than the earlier, capacitance-type generators and power supplies.

Some wire-cut technologies bypass the problem of electrolysis entirely by machining carbides while completely submerged in an insulating, dielectric oil (like a vertical EDM) instead of the more commonly used deionized water. While the insulating medium of oil does prevent oxidation and electrolysis of the workpiece, the tradeoff for this is that a very hard and brittle layer of remelted tungsten and cobalt up to 20 µm thick will remain on the machined surface. This layer is very highly stressed and can cause tension cracking of the surface or propagate microcracks into the parent material. In addition to possibly damaging the surface finish, cutting speeds in oil dielectrics are very slow when compared to water.

Special generator circuitry for carbides has recently been developed. It provides profitable cutting speeds by using water as a dielectric, yet eliminates oxidation and electrolysis, a significant contributor to cobalt depletion when wire-cutting carbide. This technology uses high frequency ac current that is delivered to an electrically isolated workpiece. This special circuitry, using alternating current (polarity), eliminates the damaging "deplating" or leaching of cobalt from carbide workpieces. This provides not only a superior part or workpiece structurally, but also a much better surface finish, which is what we were after in the first place.

SUMMARY

Surface finish and surface integrity are more than just closely related. They are an integral part of each other, and determine to a great extent whether a part or workpiece is good, acceptable, or just plain scrap.

Understanding the EDM process and how various materials react to it is the single biggest key to EDM success, especially when it comes to finish and surface integrity. The more one learns about surface finish and surface integrity, the quicker one realizes that the two are inseparable.

There is a lot more to "getting a good finish" than merely knowing what knob to turn or what button to push. Some materials are more difficult to finish than others, and some are practically impossible. As you learn the differences in approach and behavior, store them all in your memory banks and use them for future reference. Your own experience is your best resource as we all continue to *"learn to burn."*

Chapter 18

MACHINE INTELLIGENCE AND EDM

INTRODUCTION

EDM use in North America continues to advance in both sales volume and manufacturing applications. With the increase in diverse applications and expansion into many more fields than ever before, EDM technology likewise has had to grow to keep pace with the challenges that the diversity these new applications present. While our knowledge and use of the EDM process continues to grow, the machines too are also getting "smarter." This increase in machine intelligence is becoming absolutely necessary as more and more EDM operations require untended operations and machine autonomy for economic and delivery requirements.

SERVO CONTROL

During World War II, two Russian scientists — B.R. and N.I. Lazarenko — fitted a crude EDM generator with a simple, cam-driven, mechanical device that allowed the advance or retraction of the electrode in a crude attempt to maintain the proper spark gap.

As time went by and other countries took up the development of this new process, mechanical drives were abandoned for more flexible ones — either hydraulic or electric. Initially, hydraulic servos were prevalent. Feedback from the spark gap went to the control which would determine if the electrode was too close or too far away. This signal was sent to a transducer which in turn actuated a hydraulic cylinder via valving and equilibrators to advance or retract the quill of the machine. This worked satisfactorily for many years, although compared with today's standards, it was quite slow.

The eventual change to electric motors for electrode movement provided quicker response times as all signals remained electric instead of changing from electric to mechanical. Despite this improvement, all such systems were very limited in providing safe machining conditions.

ADAPTIVE CONTROLS

EDM continues to develop as these challenges continue to become more difficult and demanding. Originating from the first controls, which were "deaf and dumb" by today's standards, they progressed to what is called "adaptive" controls — not really "smarter" but perhaps more "aware." Adaptive controls are found on almost all machines today, vertical and wire alike.

The self-descriptive terminology "adaptive" describes how this process works: the control "adapts" certain parameters of cutting conditions to the ever-changing conditions encountered while EDMing a given cavity or detail.

For example, as an electrode gradually advances deeper into the workpiece, cutting conditions can change significantly. Flushing becomes increasingly difficult, the electrode and workpiece are warmer, electrode surface area has increased, etc. All of these things can add up to potentially major problems ranging from a drop-off in machining speed to disastrous DC arcs or an EDM "meltdown."

Adaptive controls operate via feedback from the spark gap, based upon criteria such as servo-voltage, gap contamination, dielectric reionization, etc. A common method of "adapting," depending upon the make, model, and age of the EDM control is to increase the servo-voltage. This will increase the distance between the electrode and the workpiece, effectively increasing the voltage and decreasing the amperage. This will slow machining operations slightly but will provide slightly better flushing due to the larger spark gap.

Some machines have the capability of increasing the increment of preset off-time. This too will slow machining rates but will provide safer machining conditions because the dielectric has more time to recover or reionize. If machining stability is recovered, the off-time will gradually be reduced to the original, preset condition, regaining efficiency.

Another feature of some adaptive controls is the automatic modification of the auto-jump or "peck cycle." If, after changing other parameters, the machine is still encountering difficulty, the control can introduce a one-time jump cycle or implement a peck cycle at random intervals as needed. This process can significantly reduce the possibility of DC arcs by aiding reionization (flushing) via hydraulics — exchanging and ejecting the oil like a piston in a cylinder.

Without question, adaptive controls have aided in the successful production of EDMed parts and have been one of the primary reasons for the increase in safety and the growing popularity of EDM. However, while providing safer machining conditions, all manipulations of cutting conditions via adaptive controls will also slow down the process, as they have no provisions to increase machining *speed*, only machining *safety*.

FUZZY LOGIC

Fuzzy logic has it origins in the 1930's, but the concept seemed so abstract that it was snubbed into obscurity by conventional-thinking science. Then, in 1965, Dr. Lotfi Zadeh, a Professor of Computer Science at the University of California, Berkeley, published a paper titled "Fuzzy Sets." He and his paper are credited for initiating the concept and development of modern fuzzy logic.

Fuzzy logic is the next step toward machine intelligence. Several EDM builders already utilize this technology to varying degrees. But, first of all, what *is* fuzzy logic?

Fuzzy logic is the terminology given to the process that allows the control unit to "think" its way "by degree," through constantly changing conditions, based upon predefined models or goals, enabling it to compensate and adjust numerous parameters to achieve the desired results. This allows the system to implement the safest, most efficient method for job completion without requiring operator intervention.

Devices or systems requiring fuzzy logic capabilities are typically found in an environment that tends to be complex or requires high-speed decision making and actuation, or both. Fuzzy logic controls can be found in equipment or systems that must make evaluations and make changes to preselected operating criteria. Examples of this can be: automatic cameras and camcorders, HVAC systems, automobile subsystems such as fuel injection units, ABS (Anti-lock Braking Systems), cruise-controls, and automatic transmissions. Fuzzy logic-controlled devices have been in use for several years, and it is only logical that an enterprise as varied and complex as EDM would eventually utilize it, especially within the CNC arena.

But why is this necessary? Why would we want fuzzy logic in EDM? Isn't EDM complex enough without adding yet another science?

Ironically, by adding another mode of computing to our generator and control, we can simplify the system we already have — one that monitors and adjusts the complexities of the primary system for us faster and with greater precision than any human. I'll explain.

As a rule, the greater the number of variables that are involved, the greater the need for fuzzy logic. In EDM we always have a large number of constantly changing conditions; therefore, fuzzy logic is applicable. Despite the need for fuzzy logic in complex operations, I'll use some very simple situations for comparison.

When driving a car, we all know that we must stop for a red traffic light and go on a green one. Since we cannot instantly stop as soon as the red light appears, the red light alone is obviously not enough information. We must be provided with an amber light to let us know we must "get ready to stop."

On HVAC systems, when the room temperature approaches the desired setting, fuzzy logic controls direct the air-conditioner to gradually "slow down" instead of continuing to blast the room with high volumes of ice-cold air, then abruptly shutting off.

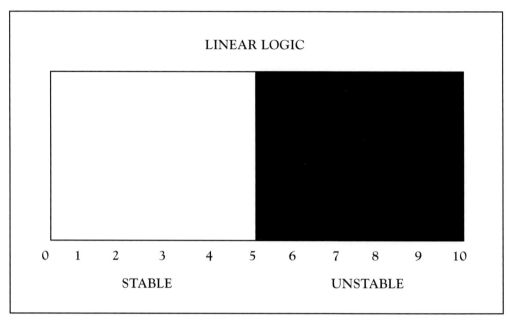

Figure 18.1   Linear logic.

Similarly, traditional computers have always operated only within the sharply defined world of switching "on or off," "yes or no," and "open or closed," making simple "black or white" decisions based upon the "linear recognition" of microchips and transistors. But for many situations, like the traffic light, this "on/off," "yes/no" criteria is no longer sufficient. Just as we need the anticipatory amber traffic light in between the red and green lights, we must also provide a "gray area" within the "black or white" world of computers — non-linear recognition, or fuzzy logic.

As shown in Figure 18.1, with simple linear logic, there is no "shading" or evaluation by degree. The world of linear logic is rigid and inflexible. So if we subscribed to everything as being either black or white, then we would be prevented from providing even the most simplistic of provisions like an amber light at a traffic signal. Using Figure 18.1 and substituting the colors red and green for stable and unstable, let's try to sort apples by color. It is usually obvious when an apple is red or green, but using this scale, what would you do with one that is "almost" red or "kind of green?" What if the chart measured temperature? If the left side is "cold" and the right side is "hot," then where would "warm" fall? How about "cool?" Is "just right" exactly on "5?"

Bringing us back to EDM, is "5" always the breakpoint for conditions being considered either stable or unstable? Does this mean that anything higher than a "5" is considered an unstable condition? Would the machine automatically shut off at "5.5" to protect the job from a possible dc arc? Where would "sort of stable" fall on Figure 18.1? Obviously we need some gray areas here.

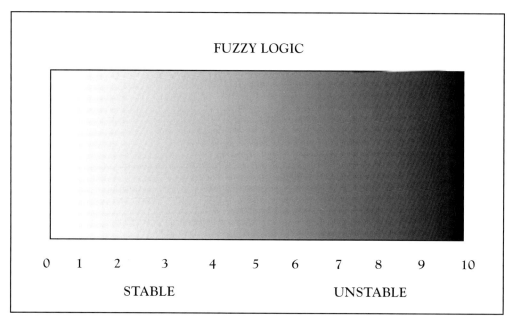

FUZZY LOGIC

0   1   2   3   4   5   6   7   8   9   10

STABLE   UNSTABLE

Figure 18.2   Fuzzy logic.

As shown in Figure 18.2, with fuzzy logic, there is "shading" or evaluation "by degree." With fuzzy logic, we can now describe apples that are "not quite red" or "almost green" (although I wouldn't recommend anyone to use either of these terms to explain the disputed color of a traffic light to a police officer!).

The fuzzy logic decision-making process allows our apples to be "almost red," "light green," or "kind of small." Our room temperature can be "kind of hot," or "just right". Our EDM machine can continue machining despite encountering occasional periods of instability. With fuzzy logic we can define machining stability "by degree."

Let's see how this process actually works. Let's control a room's temperature using a simple fuzzy system.

Rule Sets

Fuzzy logic will base its decisions and actions upon preprogrammed "rulesets" that are entered into the computer before running the program. A "ruleset" is exactly that — a "set of rules." Our air-conditioning example requires only one ruleset containing only two rules — one for temperature and a second for cooling fan speed. We would describe temperatures as shown in Figure 18.3. We would then set fan speeds as in Figure 18.4.

The temperature and fan speed rules shown in Figures 18.3 and 18.4 are called "rule sets," and they will interact to arrive quickly, efficiently, and *intelligently* at the preset room temperature.

By the "if *this*, then *that*" interrelation of the rule sets "temperature" and "cooling fan speed" (nonlinear recognition), our air-conditioner "knows" when to in-

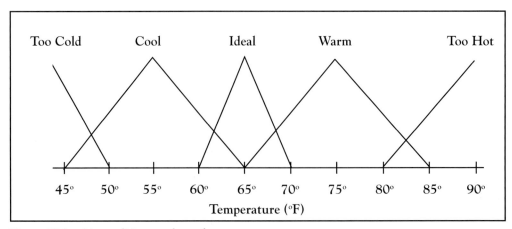

Figure 18.3   Air conditioner rule set for temperature.

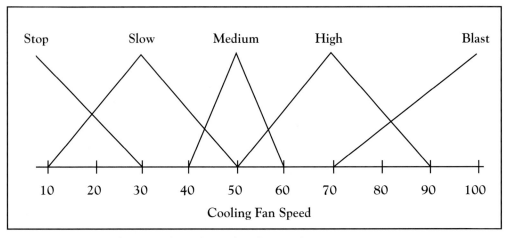

Figure 18.4   Air conditioner rule set for cooling fan speeds.

crease or decrease its output *gradually*, instead of the imprecise overshoot/under-shoot behavior typical of an simple on/off thermostat (linear recognition system).

As you can see in Figure 18.5, if the room temperature is "too hot" when we first turn on the air-conditioner, higher cooling fan speeds will quickly cool the room then gradually slow down to a more cost-efficient and comfortable speed. (For this example, only cooling criteria were used, but for a year-round HVAC system, we would also have corresponding rule sets for heating also.)

Since we are EDMers and not air-conditioning specialists, our rule sets would be EDM specific and already be resident within the CPU. These rule sets would en-compass rules regarding things such as: electrode materials, workpiece materials, flushing conditions, gap contamination, spark frequencies, etc. To execute an EDM detail, we will construct the model (set the thermostat) of our finished product. We will tell the machine certain facts about our EDM job: "It's a rib-burn" could be one. "I have multiple electrodes" could be another. "Flushing will be poor," and so on. From these data, the fuzzy-logic control will access and apply the correspond-ing rule sets much faster than any operator could.

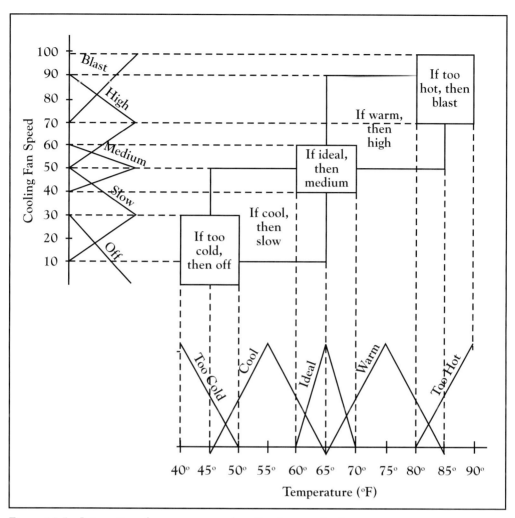

Figure 18.5   Interaction of temperature and cooling fan speed rule sets.

A fuzzy system can store just a few rules (as in these examples) or thousands of them, and they can all be interconnected for instantaneous decision making. This makes for a very precise, high-speed system of nonlinear logic, and it is far from computer "guessing" as has been implied by its detractors and reluctant believers.

A NEW MENTALITY

The old saying, "A computer does what you tell it to do, not what you want it to do," does not always apply any longer, at least when fuzzy logic is involved. Fuzzy logic can, within limits, override what you *tell* it to do and it will do what you *want* it to do!

In EDM, if an inexperienced operator enters marginal parameters into a program, fuzzy logic can examine the human-desired results (the model) and compare this against what the human entered as a program. By knowing the "rules" or model, fuzzy logic can almost always optimize and improve upon a human-generated program.

For discussion's sake, let's say we have input a setting of "5" for some EDM parameter. With conventional adaptive controls, the value of "5" for a given setting will be supplied as a simple "5," even though it may require a combination of settings *totaling* "5." This is normal binary logic behavior — straightforward, logical, and unfortunately rigid — often unable to select optimum settings which can increase job cost and the potential for mistakes.

With the enhancement of fuzzy logic, the computer will examine all of the other cutting parameters and compare their "signatures" against the preprogrammed model of "5." This allows the fuzzy logic control to supply this number "5" in combinations of 5+0, 4+1, 3+2, 2+3, 1+4, or 0+5 — ultimately supplying the setting of "5" that we initially asked for but it is supplied in the combination that is most advantageous to the success of the operation.

We have just examined how fuzzy logic can control one EDM setting, but fuzzy logic isn't designed to maintain single settings or single tasks. Fuzzy logic is best suited to monitor and adjust many different control parameters by simultaneously comparing existing conditions of each parameter against a pre set, master model to achieve the best speed, wear, finish, and stability according to the model.

Fuzzy logic supplies parallel or "ladder logic" to a decision-making process. For discussion's sake, ladder logic is similar to the "ladder charts" found in service manuals that are used to test and troubleshoot mechanical and electrical components or computer programs. Within a computer, it operates very much like a typical Basic language computer program, such as: *If* this, *then* that, *go to*, etc. Fuzzy logic can execute and maintain this entire ladder chart in milliseconds while examining its simultaneous interaction with many other "ladders" that are operating parallel to it.

For example, "If 5 = 4 + 1, then change the servo-voltage to 65 V, but if 5 = 2 + 3, then set the servo-voltage to 72.5 V, unless there is gap contamination, then increase the off-time to 45 μ seconds. If this remains stable for 500 μ seconds, then reduce the servo-voltage to...." Whew! With all of this crisscrossing of almost infinitely variable cutting conditions, one can easily see where the powerful number-crunching capability of a high-speed microprocessor is superior to the comparatively slow-motion reasoning and reaction-time of a human.

OK. But what does all of this mean? What will all this high-speed manipulation of numbers and settings provide its users?

Fuzzy logic provides an "expert system" that will narrow the gap in the performance and results obtained between skilled craftsmen and entry-level users. Fuzzy logic, in effect, supplies the experience and knowledge of an experienced user, and can implement this information much faster than any operator. The results will be reduced learning curve, increased production, almost total elimination of reject parts, reduced use of multiple electrodes, and significantly reduced operator involvement. An EDM equipped with fuzzy logic provides all of this, in addition to yielding more uniform and predictable results from operator to operator regardless of experience.

ARTIFICIAL INTELLIGENCE

Ultimately, based upon the earlier technologies of adaptive controls and present-day fuzzy logic, in the very near future most EDM's will come equipped with Artificial Intelligence (AI). Many already do. AI is considered by many to be the ultimate or "final translator" to deliver all of the information and experience that we have gleaned through the years. After learning about fuzzy systems, I personally hope this is not the case. Why? Because simply put, AI units merely employ "look-up" tables consisting of predetermined database information that is literally "looked-up" during operation and used as a parameter. Fuzzy logic needs no tables as it can arrive at a decision by "thinking" of the answer on its own.

AI operates as a linear-recognition system — a computer. Within the rigid on/off, black/white world of a computer there are no gray areas and no fuzzy. This will be good but not the best. It does learn, but only in binary logic.

Do you remember the super-computer "HAL 9000" in the movie *2001?* HAL stood for: <u>H</u>euristically programmed, <u>AL</u>gorithmic Computer. (Heuristic is "a problem-solving technique," and Algorithmic is "a mathematical rule or procedure for solving a problem.")

HAL used numbers to solve problems. Computer numbers like 0's and 1's were his binary language, recognizing only "on or off," "yes or no," and/or "black or white." Being a linear recognition system, HAL is an example of Artificial Intelligence. HAL tried to serve man based upon rigid preprogrammed mathematical rules and constants, but he had no provision to make decisions based upon the numerous gray areas and variables of human behavior.

Since he was a "black and white" computer, he could only act in "black and white" and after tragic circumstances, he had to be disconnected because gray areas of information invaded and confused his black-and-white world of computer logic. Because he was a linear recognition system, he was unable to evaluate things "by degree." HAL's binary programming told him that humans either lie or they tell the truth. He did not know that the truth (about the mission) can be told "in degrees." When he became "confused" by "degrees of truth," he did what he was programmed to do by default — to complete the mission, even if it meant expending the humans.

This example certainly does not mean that every AI unit will ultimately fail in their missions or destroy their human creators. But this example of HAL does point out the potential limitations of linear recognition in a high-end system.

For some systems this will be fine — robots sorting fasteners, inspecting machined parts, matching paint colors etc. — but most high-end systems encountering ever-changing variables will need the ability to use nonlinear recognition to achieve the best results. The fuzzier, the better. That brings us to the next level — neural networks.

NEURAL NETWORKS

The only limitation a simple fuzzy system has is that it is an inanimate system. It can make complex decisions at very high speeds, but it cannot retain memory. Simple fuzzy systems "learn" as they run and get more efficient the longer they gather information, but they "forget" everything when it stops.

For example, with a limited or simple fuzzy system, we can successfully EDM an eight-hour mold-cavity in January, and then EDM the same mold again the following June and the system will follow the same learning curve on both occasions. It does not "remember" what it did the last time. The next logical step is neural networks. Without getting too involved with science and physics (or too far away from EDM), I will forego explaining exactly what a neural network is (it would take a book of its own) and go on to briefly explain how it works.

A neural network is, by definition: "neural," as in neuron or nerve; and "network," as in system or infrastructure.

Simply put, a neural network acts as a nervous system or, better yet, a brain. A neural network is not actually a brain, of course, but it does emulate the way a brain thinks and learns. A neural net learns by example or experience. In operation, the more information or experience it gathers, the smarter it gets.

Let's allow our neural net to redo the January/June mold cavity discussed above. Neural net units will start the job in June *at the same level of expertise it left off with in January*. Even as it progresses with the June job, it will continue to learn as it works, constantly improving and adding information to what it has already learned, surpassing its own performance in January while continually reducing all chance of errors and *never making the same mistake twice*.

Like fuzzy logic, neural nets will be used for a great number of things including EDM. A neural net will have the ability to recognize whether your signature is authentic, perform critical medical testing, or analyze rock samples from the moon. Where the computer within a simple fuzzy system must initially learn from an expert that has programmed its knowledge base, neural nets learn by actual experience. Instead of a programmer teaching it fuzzy rules before it is put into production, this adaptive fuzzy system has the ability to "see" and recognize examples it encounters *during* production. From these examples, it would constantly sort and categorize the patterns and trends in machine performance, *creating its own* fuzzy rules as it gains experience. It *learns*.

> "*A neural network acts as an associative memory*
> *that stores and compares associations*
> *between inputs and outputs, stimuli and responses.*" *

---

* Kosko, B. Ph.D. *Fuzzy Thinking – The New Science of Fuzzy Logic.* New York: Hyperion, 1993.

In theory, if *this* happens, then it (the neural net) does *this*. Depending upon what *this* does, the neural net will do *that*, etc. And it *remembers*.

In practice, if gap voltage drops below 35 V, the system will increase it by 15%, raising it to 40 V. Although conditions are better, machining speed will drop because of the resulting decrease in amperage. Therefore, the amperage will be incremented slightly to compensate. Because the higher amperage will have a slightly larger overcut, the orbiting steps of the finishing cycles must be adjusted, etc.

After performing all these calculations and permutations, this knowledge will be stored and used for future reference, constantly improving upon its own performance.

Perhaps the best of these worlds (EDM especially) would be to integrate a fuzzy system with neural network capabilities, creating an "adaptive fuzzy system." I am confident that at the rate technology is developing, all of this will come to pass within the next ten years or so.

In only ten years? So soon? Certainly!! Why not? All of this awesome technology already exists! It is only a matter of time (and economics) before features like fuzzy logic, neural networks, and adaptive fuzzy systems reside in almost all complex CPU-driven machines, EDM included. EDM *especially!*

CONCLUSION

In this chapter, we have examined where we *were* (the Lazarenkos, mechanical servo-systems, etc.), and where we are *now* (adaptive controls, fuzzy logic, etc.). So the logical (fuzzy logical?) question remaining is,

"Where are we going?" (I thought you'd never ask!)

The answer is: **To our collective, manufacturing <u>Future</u>, that's where!**

THE FUTURE

American manufacturing in particular, and international manufacturing in general, will depend more and more upon intelligent machines. As the economies of Europe, Mexico, and the Pacific Rim continue to develop, competition with these emerging nations will also increase. Intelligent machines will provide all manufacturing disciplines with the electronic equivalent of

1) knowledge,
2) skill,
3) experience, and
4) reliability.

1) The *knowledge* of experts: the comprehensive library of past knowledge that can be stored and can be accessed instantly to solve present and future problems.

2) The *skill* of a craftsman: the pool of skilled and qualified tradespeople is shrinking rapidly; machine intelligence will empower a less skilled person to perform like one.

3) The *experience* of a journeyman: the provision of literally *centuries* of collective experiences will be in the form of a microchip and used to make the best decision.

4) The *reliability* of machines: the proven dependability of modern machinery regardless of task or environment.

Ok, EDMers, now what? You may be asking, "Now that I know more about machine intelligence than I perhaps cared to, what does all of this "new science" mean to me and my EDM and my shop? How will it help me? How will it help my business? *How will it help me make more money?*"

These are good questions. Intelligent EDM's will help us accomplish a myriad of things.

1) Reduce operator training time and learning curves. This will also directly reduce scrap and rework substantially.

2) Immediately infuse years of experience and machine technology into the machining and manufacturing trades to offset the dwindling supply of skilled craftsmen.

3) Produce EDMed parts with good finishes and minimal recast — quickly and safely — regardless of workpiece material.

4) Expand the productivity of all shops as machine intelligence increasingly supports untended operations and machine tool autonomy.

5) Quicker deliveries/faster turnaround times. Because of untended operations, more work will be produced in less time, shortening job lead time.

6) Increase process control and predictability. Both of these have been a long-perceived "black art" when accurately quoting jobs and predicting results.

7) Enable small companies to better compete with the larger ones who traditionally have been more capable, further broadening our manufacturing base.

8) Create output parity. Intelligent EDM's will help produce consistent results from machine to machine and man to man, regardless of operator experience.

Increase our manufacturing diversity and capabilities. Because we will no longer have to "reinvent the wheel" on every job, we will have more time and resources to attempt the unknown.

SUMMARY

There are many more benefits to intelligent EDM's, but I believe I've made my point. Competing in the '90's has been difficult enough, and it will only get tougher as we pass into the new century and meet the challenges that the new, ever changing global economy will present. The success or failure of our manufacturing future, especially where EDM is concerned, will depend to a large degree on whether we embrace this new and slightly daunting technology or resist it.

To resist the upcoming and inevitable changes in the machine-control environment could very well prove fatal to our manufacturing future. We must make adjustments. We must evolve, we must *adapt* to changing conditions or face the consequences. Remember the dinosaurs? Dinosaurs couldn't or wouldn't adapt and we don't see them around anywhere, do we? This is exactly my point:

We don't need any "Jurassic Sparks" in EDM, either.

But maybe *adapt*ing isn't quite enough. Eventually, adaptive controls will become technologically inadequate to meet our ever-demanding EDM needs. Perhaps being only adaptive-minded in our thoughts about improved manufacturing strategies might not be enough either. Instead of merely adapting to a new situation or process, perhaps we should actually strive to be *part of its development*.

Think about this for a moment: in order to be adaptive, to react, *something must have happened <u>first</u>*.

In the light-speed world of EDM, this reaction could very well come too late. (We work in time-units of *micro* and *nano*seconds, remember?) To stay current and perhaps get ahead in this business, we must evolve and become *proactive* and not *reactive*. With the new-found knowledge and experience of computer intelligence, tomorrow's EDM's will be proactive by basing all of their new decisions on ones made previously. (You know, kind of like the way humans are *supposed* to work).

As for the dinosaurs — we have a significant advantage over them in that we have vision. Not vision as in eyesight, but vision as in *foresight*. We can see all these things coming. We know our manufacturing world is rapidly evolving. This evolution is ongoing and will not wait for us to "get ready" for it. We must be ready to meet it head-on or it will pass us by just as quickly as it is evolving.

All those who don't believe in or won't participate in evolution are most certainly destined for extinction. And if you have any doubts at all about this, just ask any dinosaur.

Chapter 19

EMPLOYMENT OPPORTUNITIES IN EDM

INTRODUCTION

Electrical Discharge Machining is one of the oldest "emerging technologies" in existence. From its first primitive days of tap busting to the refined finesse of today's 32-bit, fuzzy logic CNC units, EDM has provided an ever-increasing base of employment opportunities in shops across the country. Along with the obvious EDM-related job opportunities in manufacturing, there are a lot of "not-so-obvious" ones also. We are going to examine some of them.

Before we look at some of the specific titles and job descriptions, we can look briefly at some of the fields and industries that EDM has influenced. Within these fields you will find some of the EDM-related jobs that follow this list.

FIELDS OF OPPORTUNITY

TOOL AND DIE

Blanking	Coining
Extrusion	Forming
Piercing	Swaging
Bending	Drawing
Fine Blanking	Heading
Progressive	Trim

MOLDMAKING

Blow	Die Castings
Injection	Reactions
Rubber	Transfer
Compression	Forgings
Investment	Rotation
Thermoset	Vacuum Form

PRODUCTION

Aerospace	Automotive
Dental	Hydraulic
Medical	Optical

JOB SHOP

2nd Operation	Gauge Making
Hydraulics	Masters
Short Run	Tool and Die

Aircraft	Cutting Tools	Cutting Tools	General
Electronics	Jigs/Fixtures	Jigs/Fixtures	Repair
Oil Tools	Templates	Standards	Tooling

PROTOTYPING		**RESEARCH AND DEVELOPMENT**	
Cutting Tools	Gauge Making	Aerospace	Ceramics
Jigs/Fixtures	Masters	Electronics	Optical
Patterns	Standards	Metallurgy	Nuclear
Development	Investment	Aircraft	Dental
Loft Tooling	Model Making	New Materials	Medical
Short Run	Testing	Military	Rockerty

There are many, many other categories and job descriptions — far too many to list here. These are only a few areas of EDM influence but they should provide an overview of what to expect when one is examining EDM's impact upon modern manufacturing.

The following is a listing of a few EDM-related job titles and responsibilities.

OPERATORS

This is where it typically all begins — in front of an EDM machine. Unfortunately, all but a very few of us ever received any formal training, and perhaps only from an associate who likewise, learned by the seat of *his* pants. Unfortunately, for the most part, most operators over the age of 35 are self-taught.

From the standpoint of any operation's profitability and earning power, this is a worst-case scenario — an operator (expensive) learning EDM the hard way (expensive) and making mistakes on payroll parts (expensive). Perhaps, eventually, this will prove to be a great advantage *if* the shop can survive long enough to reap the rewards. The operator will have become proficient, only because after several months or years, he will have learned not necessarily *what* to do but what *not* to do.

Fortunately, through the greater acceptance of EDM by manufacturing, sales have increased steadily; and with the larger numbers of EDM machines being installed, more and more people are being properly trained every day.

As this chapter develops, you will see and understand why *training* is the single most important facet of EDM. Experience is valuable too, but not everyone has it, and it is difficult to come by without receiving training in the first place!

The requirement of possessing a good, working knowledge of EDM is enmeshed throughout manufacturing, touching every discipline from design, process, and manufacturing engineers to draftsmen, programmers, operators and managers. It also encompasses salesmen, service engineers, applications engineers, and the suppliers of EDM raw materials, tooling and consumables.

Note: If you are not knowledgeable in this craft, you are doomed to failure. Some failures will prove prolonged and painful, others can be more quickly, but not necessarily without pain.

Whether it is wire-cut or vertical, we must *"Learn to Burn."*

TOOLMAKER/MOLDMAKER

Journeyman is the title or job description attained after serving a formal apprenticeship program, usually four years, and is the equivalent of an Associate of Arts (AA) degree. There are state-run programs that are sponsored in conjunction with companies who cooperate within training guidelines. For these companies, in return for helping the state educate and train badly needed, skilled craftsmen, special considerations and certain tax benefits are available.

The journeyman must be keen of eye, steady of hand, sharp in mathematics, and have an even temperament if he is to be successful at toolmaking, especially if he expects to continue to advance upwards into management. These abilities, plus the added value of knowledge and skill in EDM, can place this individual at the peak of desirability in any toolroom.

The journeyman will command top wages for his craft and rightfully so. You must consider that day after day he must strive for perfection — *that is his job*. To provide anything less is *not* doing his job. Therefore, in theory, he can never *exceed* expectations, he can only *meet* them or *fall short* of them. While these kinds of pressures and demands can be tremendous, the rewards and benefits can likewise be proportional and equitable.

LEAD MAN/FOREMAN

The next logical step for an individual in a shop or toolroom would be running or managing the shop operations. Since EDM is one of the most difficult crafts to master and is typically augmented by proficiency with most other shop equipment, EDM experience can be the deciding factor for promotion. Compared to the number of lathe and mill specialists in manufacturing, there are very few "EDM experts," therefore, they are at a premium and will command the wages and prestige that comes along with "dues paid."

Barring sometimes unavoidable personal or political situations, this move is typically based not only upon an individual's shop ability and aptitude, but an attribute totally unrelated to metalworking — *social skills*. This is one part of toolroom etiquette that, unfortunately for some individuals, can be very difficult to learn and accept, but is absolutely necessary in order to make any meaningful advancement as this is a logical and obvious first step into management.

At this point, we can break away from actual machine operations as not all EDM jobs are in the shop. The EDM related jobs that follow provide multiple opportunities for equally secure and profitable peripheral and support positions.

PROGRAMMERS

Most wire-cut jobs require complex programs for contouring and/or tapering and, therefore, require skilled and knowledgeable people to devise them. While in this position, one may find oneself using, testing, or specifying the very latest in 4-axes CAD/CAM programming systems. This will also keep one abreast of the very latest in computer hardware also.

Although it is not absolutely necessary, having previous experience operating a wire machine is a definite plus. By having operated one, it will provide a much greater insight to programming strategies such as strategic placement of start holes and cutoff tabs, "nesting" of parts for efficient use of material, and material behaviors such as stress movement and finish requirements. This experience is invaluable and will help secure both the position and the compensation consistent with its responsibility.

DESIGNERS

In the past, most EDMed parts were either cores and cavities for molds, or punches and die components for the die-making industries. This is not so any longer. With EDM reaching into almost every facet of modern manufacturing, we often have to rethink how we look at part and product design — with EDM in mind, *beforehand*.

Again, having previous skill operating an EDM machine is a definite plus, and in this position, perhaps almost mandatory. One must have a thorough, working knowledge of machine tool capability before being able to exploit it. Having run one will enable a parts designer to "see" or think of a design in an unconventional or nontraditional manner. Since EDM is still regarded by many as an unconventional or nontraditional process, then why shouldn't our thought processes be likewise? Design with the EDM process in mind, instead of it being an afterthought. This same train of thought could be transferred to the following job positions:

MANUFACTURING ENGINEERS

These are the people who, after the part is designed and approved for production, must undertake an intensive examination of all possible methods of manufacture for feasibility and cost-effectiveness. Those knowledgeable in EDM can, in many cases, manufacture that "impossible" part or eliminate several conventional machining or postprocessing operations by specifying EDM in the part's route sheet or traveler. Whether in the design stages or in manufacturing engineering, the individual with a broad knowledge of EDM and applications can have a significant advantage over those who do not.

APPLICATIONS ENGINEERS

For those shop people who seek to get "off the bench" in a traditional sense, this can be a logical and lucrative pursuit. This can be done most typically for a major builder of EDM or, in some cases, for a larger EDM distributor or dealer.

Utilizing one's existing knowledge and shop experience, a successful candidate for this position will have the opportunity to learn and use some of the latest advancements in equipment, technology, tooling, and supplies, while working with some of the most diverse and sometimes demanding situations and applications.

Test cuts and feasibility studies for prospective customers will provide the highest degree of challenge to those who are up to it. Challenges include unconventional applications, new and exotic alloys, the intrigue of working on sensitive or proprietary projects that will require nondisclosure statements or security checks, not to mention the personal challenge of competing against "the other engineer over at 'Brand-X' Machinery."

This position can also offer the candidate the opportunity to devise, set up, and run some of the demonstrations shown during high-profile machinery tool shows and exhibits, which can mean travel to different parts of the country and meeting some interesting and exciting people from all phases of manufacturing.

SERVICE ENGINEERS

For those technically inclined and with electrical or trade-school experience, this can be another career option. It offers exposure to some of the more dynamic and high-end people and shops in the country. Perhaps, to some, this is not necessarily considered a "glamorous" position, but it is a key position nevertheless, and those who pursue this discipline have the satisfaction of being called upon to solve sometimes complex and challenging problems on the customer's floor. Many times, the pressure can be intense as deliveries and contracts may hang in the balance of the performance of the machine.

Successful service engineers should enjoy steady, secure employment and regular increases in compensation as they are an integral part of a company's service and support team.

INSTRUCTORS AND TRAINERS

As cited early on in this chapter, very few of us have enjoyed the luxury of having had formal EDM training — at least among people *my* age. Fortunately, with the increased number of units being sold, more and more people are being properly trained in both EDM theory and practice. This is good for the EDM industry in particular and American manufacturing as a whole, as successful customers will continue to buy and recommend EDM as a viable and profitable enterprise, thus helping expand the entire industry.

Those with not only the gift of knowledge where EDM is concerned, but who possess the ability and need to help people by teaching them to help themselves, will find this position not only challenging but immensely gratifying also.

Compensation in salary and satisfaction can be very attractive and job security is very good as EDM sales continue to increase and therefore, so does the need for EDM training.

SALES

This just seems to be a natural place for many people to wind up who have been involved in EDM — selling the process and product that they have been using for years. Why not? After using a certain brand or type of machine for years, who could be better qualified to promote its merits and abilities? Granted, not all EDM people have the drive or social skills required to be successful in sales, but those who do have an incredible opportunity to advance their careers and earning potential.

For those who are well versed in EDM operations but are reluctant to consider the leap into the world of sales, consider this: "It is usually much easier to teach *sales* to an EDM person than it is to teach *EDM* to a salesperson." This does not apply in every case, but it is more the rule than the exception.

This observation is based upon the fact that EDM is not a brochure-sold commodity. It is a very technical sale, and the potential customer will ask some very demanding technical questions, many of them related to the *process* and not always just the *product*. The salesperson with the ability to field these questions readily will have a greater advantage over an individual who promises, "Uh, I'll have to get back to you on that one." The salesperson with "chips in his shoes" will have much more credibility than the "brochure salesman".

At this point, sales positions can branch off into many different directions — all of them potential career positions with security and success based upon product knowledge.

SUPPLIERS

Numerous sales opportunities exist on the supply side of EDM. This encompasses everything from raw materials, consumables, EDM tooling, filtration systems, and the sale of other EDM related products.

Raw materials would include copper and graphite electrode blanks, copper and brass and tungsten tubing, copper tungsten and silver tungsten blanks, etc. EDM consumables would include: filters, resin, wire, diamond guides, carbide contacts, standard and high-performance flush cups and nozzles, dielectric oils, fire-prevention devices, etc.

EDM tooling is an industry unto itself, offering sales and engineering positions dedicated strictly to the successful integration of work and the machine tool. EDM tooling encompasses many different styles of workholding devices, including those for setup and inspection, rotation and C-axis units, automatic toolchangers, automatic workpiece changers, palletization, and robotics. The diverse field of EDM tooling offers sales, service, and engineering opportunities.

Likewise, the filtration and water-treatment industries undergo continuous new developments to provide numerous possibilities of potential employment. Whether your interests lie in design, manufacture, sales, or service, opportunities abound in this fast-growing and highly competitive field.

With the advent of increased awareness and sensitivity of the environment, another potential industry opens up to the EDM entrepreneur — waste disposal. Gone are the days when we can simply toss our dirty filters into the dumpster. Chlorines and heavy metals contaminate these dirty filters, and D.I. (deionized) resins contain strong acids and bases along with toxic metals and other materials and have always been restricted from "casual" disposal. Citations and fines are being issued to violators as you read this.

State and/or federally approved disposal companies and licensed carriers will soon be handling all EDM waste disposal, whether it is spent EDM oil, water, filters, or resin. All of this waste will have to be specially treated, recycled or disposed of properly. This will create jobs and revenue for a whole new facet of the industry, just as the graphite industries expanded and refined their operations to support the requirements of EDM.

CONSULTING

There is a very high demand these days for knowledgeable EDM consultants. As EDM installations become more prolific and even more diverse in their applications, so does the need for professional EDM consultants.

A cynical observation concerning consultants has been, "A consultant is someone who is merely in between jobs." Perhaps, for some, this is true, but speaking from personal knowledge and experience, there is quite a demand for experienced EDM people willing to share their expertise and experience with those who know less and, therefore, have the resulting problems.

One potential drawback one must be made aware of, especially when starting out in this field, is that a lot of promoting and marketing must be done in order to "get the word out" that your services are available. Your first few jobs will be spotty, and most potential clients will request a list of references. Once an individual consultant or consulting firm's credibility has been established, consulting opportunities abound, especially among Fortune 500 companies here and larger corporations abroad (the Pacific Rim in particular). Contracts can be quite lucrative and can be for a specific project or run for several years, especially when overseas travel is involved. EDM consulting can be a very high-paying and prestigious career for those who qualify.

Again, another field of enterprise has been created by the need for more education and training, supported by the individual's knowledge and experience in EDM.

SHOP OWNER

Here is where we could leave the subject of EDM entirely and start conducting a class on "Business 101," but we won't. Instead, here is just a brief summary of what this might entail.

For many shop people, owning their own company and being their own boss is the ultimate goal. And why not? After all, this is America, the land of opportunity and if you can't do it here, chances are good that you will not be able to do it anywhere else either.

On the surface it sounds like a wonderful idea, and a great many people have been successful and built large companies while amassing large sums of money. A few people have not only started and run very successful EDM shops, but have gone on into the fields of EDM research and development, further enhancing their own capabilities and helping to make this new technology available to the manufacturing community.

Obviously, success to this degree is the exception and not the rule as many owners are not as capable or are content to operate on a much smaller scale. Unfortunately, statistics on this are brutal and the majority of start-ups fail. Why?

To answer this question we must first examine the entrepreneur's motivation for owning his own business. Is it for independence, financial advantage, to build a better mousetrap, or perhaps a combination of all three? Sometimes employees leave under duress or in anger, determined to prove something by starting their own business. All can be valid reasons, but owning your own business is a totally different experience for anyone whose experience is restricted to the less volatile world of payroll jobs.

Before starting out on your own, you must already believe in yourself and have confidence in your skills as a craftsman. This faith or confidence is not only good, it is absolutely necessary. However, the fact remains that while you might very well be the world's greatest EDM specialist or toolmaker, that doesn't necessarily make you the world's greatest businessman. There is much more to being a shop owner than just "making chips." Remember that over 90% of all small businesses fail in their first two years.

You should have confirmed purchase orders, *in writing*, before hanging out your sign. This presents something of a paradox because *how can you get work if you aren't in business yet?*. If you have the good fortune to leave your payroll employer with good relations, they can be an excellent source of steady work that you are already accustomed to. A great situation.

Otherwise, you are left to procure work. Which means that while you are out selling yourself and your services, you aren't in the shop to answer the phone for that million dollar contract that is bound to come in when you are out. A phone machine? It's better than nothing, but most callers really dislike these things when trying to do business. You are much better off hiring a person or service to take your calls, but that will increase your overhead.

Another situation facing the new, one man shop is selling versus machining. While you are out soliciting work, you aren't cutting (unattended EDM being an exception).

There are numerous areas in running a business that most people aren't aware of until they are already committed. Sure, you know all about buying the right machine and you already have your own tools, but have you considered licensing, leasing, power, lighting, phones, faxes, stationary, insurance, taxes and don't forget the billing cycle. Even though a bill typically says "Net 30 days," what actually happens can be very different. For this reason alone, you should have 3 to 6 months working capital available "just-in-case."

While on the subject of money, watch out for banks and other lending institutions. They will typically lend you just enough money to get yourself into trouble. If you think you need $50,000, ask for $100,000. They might lend you $80,000 and then you might have enough.

Please don't misunderstand my intentions here. I am not trying to talk anyone into or out of anything, but I would be remiss in my responsibilities if I did not alert readers to all aspects, good and bad, of owning and operating a small business. Many before you have prospered and enjoyed success, but many more have failed. Don't be afraid, but *be aware*.

SUMMARY

Whether you seek employment as an operator, toolmaker, foreman, manager, or engineer, or in sales, service, applications, supplies, consumables, or waste disposal, you can clearly see there are plenty of career opportunities in EDM.

Besides being part of the dynamic and exciting world of EDM, this craft provides all of the necessary ingredients of a career position, regardless of title or job description.

These are as follows:

1) **Compensation**. Skilled EDM operators are traditionally among the highest-paid wage earners in a shop environment. Since operators have learned to master a skill that management has long considered to be a "black art," shouldn't they be entitled to pay that is commensurate with their skills and experience? Logic says so.

   Further, if one has taken this EDM skill and knowledge to another level or job description, then their previous efforts and experience should likewise be taken into consideration when negotiating salaries and benefits.

2) **Benefits**. Most employers realize how difficult it is to find skilled craftsmen. They also know how important it is to not merely *keep* them, but *keep them happy*. This can be done in ways other than direct compensation to prevent them from thinking or worrying about nonwork-related matters. For instance, if a worker has sick-pay benefits and a good insurance package, he will be less inclined to worry about the high-cost of medical bills, and the need to look elsewhere for higher pay. Skilled craftsmen, sharp managers, and engineers are very valuable commodities and should be treated as such.

3) **Security**. *Fact*: In Japan, which geographically is only the size of California, there were around 7,000 CNC EDM units installed in 1993. In the same year, in all of America, there were only about 800 units in-

stalled. Observation: If we have the potential to expand our EDM industry almost 9 *times* its present size just to achieve parity with Japan (the size of California), just think about the proportionate potential growth for EDM in the other 49 other states. This would also include all other EDM-related industries, i.e., sales, suppliers, operators, engineers, etc. With room for this kind of growth and future expansion, this practically guarantees the EDM industry a very secure and stable future.

4) **Satisfaction**. Most of us spend more of our waking hours on the job than we do with our families. Unfortunately, other than winning the lottery or inheriting a fortune from an obscure rich uncle, we have few alternatives.

In order to spend this much time doing anything, we should at least *like* what we do. EDM offers such a broad spectrum of interesting applications and employment opportunities, the odds of finding your particular "niche" in EDM are quite good.

Whether you derive satisfaction by successfully machining the "impossible" part, in writing a very complex CNC program, or troubleshooting the "machine from hell," these are all very gratifying accomplishments. Since none of us gets raises for every completed task or job, we must at least be satisfied with ourselves and be able to derive enough pride and satisfaction from a job well done until the next review time.

IN CLOSING

If you haven't already done so, you might consider EDM as part of your future. You might like to operate a machine or actually own one. You may enjoy programming a wire machine or want to get into design work. You could manage an EDM shop or department, or work as an engineer for one of the builders. Instructor, consultant, supplier — the possibilities are almost limitless and all very gratifying.

If you have any doubt about pursuing a career in EDM, ask someone already working in the position you desire (not your boss, though). Ask other operators, programmers, and engineers how they feel about what they are doing. When you attend the next major tool show, make sure you examine the people working in the booths — not just the equipment. Do you belong in sales? Would you like to be an applications or service engineer? Talk to any of these people about a future in EDM. You won't be sorry.

Should you choose to explore EDM for the first time, or are broadening your horizons and expanding upon your existing EDM experience, I wish you:

"All the Best," in whatever endeavor you pursue.

Chapter 20

HOW TO SELECT AND JUSTIFY AN EDM MACHINE

INTRODUCTION

Of all the topics that I have covered in this book, this subject is unquestionably the one that I am most frequently asked about, and it is one of the most difficult to write about. Difficult because with so many different applications already being performed, and so many more to be discovered, it is very difficult to list the selection criteria that will satisfy the "average" EDM user. Come on... who are we kidding? There is no such thing as an "average" EDM user. If you are already an EDM user, do you seriously consider yourself "average?"

I think not.

Curiously, despite my obvious corporate affiliation, people still corner me to ask, "Tell me, Bud, you know all the machines out there... which machine is *really* the best?" People expect me to somehow give them greater insight than I actually have, to somehow give them an inside-track on their competition like it is some hot tip at the racetrack. "Psst... hey, it's OK. You can tell *me*."

Sorry, it's just not like that. Not like that at all. Unless it is a case of a repeat order to duplicate an already running application, *any* machine-tool selection must be a totally objective, case-by-case evaluation process based upon specific known requirements, conditions, and parameters. I know of no other method to do this, scientific or otherwise.

PART ONE, or "OK, SO HOW DO I DO THIS?"

We will examine the actual process of machine tool selection in an outline form using a step-by-step, comprehensive method that goes back to the very first time the thought of purchasing an EDM occurred to you. (By substituting the term EDM with lathe, mill, grinder, etc., and the machine's features, this chapter can be used as a template for almost *any* machine-tool selection and justification. In this case it will examine a wire EDM.)

The first thing to be done is prove whether or not this exercise is even necessary. Therefore, first of all you must establish or determine a:

NEED
1) Do I already have a need?
    a) Am I presently jobbing out my EDM work?
    b) Are my prices and deliveries influenced by sending my EDM work out?
    c) Could I keep my EDM in-house?
    d) If I did, what are my expected savings?
2) Should I consider a need?
    a) Am I presently missing any EDM work opportunities?
    b) Am I likely to miss any similar work opportunities in the future?
3) Do I want a need?
    a) How serious *am* I?

To further qualify this possible need, examine the following criteria:

WORK
1) What kind of work requiring EDM am I already doing?
    a) Size, shape and type(s) of material
    b) How much? (Dollar average.)
    c) How often? (Monthly average.)
    d) None.
2) What kind of work am I sending out?
    a) How much? (Dollar average.)
    b) How often? (Monthly average.)
    c) None.
3) What kind of work am I passing over?
    a) How much? (Dollar average.)
    b) How often? (Monthly average.)
    c) None.
4) What kind of work *could* I be doing?
    a) How much? (Dollar average.)
    b) How often? (Monthly average.)

After deciding upon the type of work you intend to pursue or focus on, you will be much better able to determine your:

BUDGET
1) How much money do I have to spend?
    a) What are the fiscal limitations of this venture?
2) How much money do I *need* to spend?
    a) If I purchase everything I need, how much will it cost?
    b) Do I have to purchase everything I need all at once?
3) What kind of terms are necessary/available to complete this transaction?
    a) Cash?
    b) Financing?
    c) Other?

Having completed the NEED, WORK, AND BUDGET sections, you can go on to:

PART TWO

If you have determined you have a need for EDM and have survived the initial budget questions, we can move onto the nuts and bolts of the project — actual machine selection. I'll try to offer a word of caution here without sounding presumptuous *but*: you had best know your *job*, your *business*, your *market* and your *competition* because if you don't, any mistakes made in the next section can very well prove fatal to your well-intended efforts.

PRIORITIZE THE SELECTION CRITERIA

Other than the physical limitations of the machine (size, travels, weight capacities, etc.), you must establish what items or features are most important to you and weigh their scores accordingly. Users who expect their machines to run unattended, around the clock will be interested in AUTONOMY. Another shop might need the utmost ACCURACY, while a production shop would be very interested in SPEED. Understand? Let's proceed.

PHYSICAL SIZE
1) Based on my expected part size and shape, how large a machine will I need?
    a) Do I need table *size* or table *travel*?
    b) Is my part size limited by a rigid tank?
    c) What is the tallest part I'm likely to encounter?
    d) What is the workpiece weight capacity of the machine?
2) Are there any size limitations of my facility for this machine? (Ceiling height, door width, access areas, etc.)

CUTTING SPEED

1) How fast do I have to produce this part to show profit?
    a) Ideally?
    b) Marginally?
    c) Realistically?
2) What type of wire (and its expense) is required to support question #1?
3) Are my desired cutting speeds compatible with my accuracy and finish requirements?

Important: Cutting speed can be misleading, so I must break away from this outline to explain how to evaluate this often confusing area of selection. I wanted to make this plan as simple as possible but this section does warrant some explanation.

Cutting speeds for wire-cut EDM continue to increase and they are typically expressed in square inches per hour or, in.²/hr. Presently, there is near parity in today's modern generators with cutting speed claims running from the mid to upper 20's, averaging around 25 in.²/hr. A machine selection based upon claims of faster cutting speeds alone may not always be the best choice. Fast cutting speed alone may not be indicative of a machine's true capability. Sometimes a slower machine is faster. I'll explain.

If you intend to wire-cut production parts with a single, high-speed pass, speed is obviously important but you must first consider other things that fast cutting speeds will impact. If, at your desired cutting speeds, all blueprint criteria are met — the parts are acceptable in finish, surface integrity and accuracy — then it's probably safe to consider the faster rated wire machine. If the faster cutting machine adversely affects other parameters of part specifications then maybe we'd better slow down.

I'll make an analogy to automobile performance to make this clearer: If you have two machines side-by-side making a straight-line cut like a drag race, then, barring breakage, the fastest car will win. If this is the kind of part machining you intend to do — simple shapes, open tolerance, single-pass machining with limited finish callout — then cutting speed is clearly more important to you. If your part requires good straightness, crisp geometry and sharp corners, then the dragster-type car may be all wrong for this road course.

Why? Because fast, high power speed settings will be less accurate and have poorer finishes than parts that have been skimmed. Again, if your part specifications aren't sensitive to this, then, ok, buy the dragster. If part geometry, straightness, and accuracy are critical, then the total number of additional passes that will be required to correct the conditions *caused* by the speed cut is also a critical factor.

Conclusion: Cutting speed will become much less of a factor when part accuracy, straightness, and finish is involved. A slower cutting machine might be selected over a faster one if it can meet finish and accuracy specifications in fewer passes or skim-cuts. There is a big difference between speed and efficiency. It's not always how fast you can finish the *cut*, it's how fast you can finish the *part*.

PART ACCURACY
1) What kinds of accuracies do my average parts call out?
   a) Straightness?
   b) Position?
   c) Geometry?
2) What kinds of *average* part accuracies can I expect?
   a) Straightness?
   b) Position?
   c) Geometry?
3) What are the *best* part accuracies I can expect?
   a) Straightness?
   b) Position?
   c) Geometry?
4) How long will it take to obtain these accuracies? (Machining time.)
5) What special provisions (if any) are required to achieve top accuracies? (Dielectric chillers, temperature-controlled environment, isolated tooling, etc.)

SURFACE FINISH AND SURFACE INTEGRITY
1) What surface finish is acceptable?
   a) What are the limitations of the machine?
   b) How long will it take to obtain the necessary finish?
2) How much influence does the EDM process have on the finished surface?
   a) Is the machine-finished surface acceptable to my customer *as is*?
   b) If not, what kind of postprocessing is required/acceptable?
   c) What is involved in postprocessing? (Time, money, risk, etc.)
   d) If postprocessing is needed and is acceptable, is the part process still profitable?

MACHINE FEATURES
1) Wet or dry?
   a) Examine the pros and cons of submerged and nonsubmerged machining.
   b) Will 80% of your work be executed satisfactorily with your choice?
2) Wire sizes.
   a) What is the smallest kerf or radius requirement on my average part?
   b) What is the machine's standard range of wire sizes?
   c) Can this range be expanded (if necessary) with options?
   d) Can all wire sizes be used in conjunction with an auto-threader?

3) Auto-threader.
    a) Are there multiple start-holes or parts that will require numerous re-threads?
    b) Will the application present a high probability of wire breakage problems?
    c) Does the job require continuous operation with minimal interruption?
    d) What percentage of time is the machine expected to operate unattended?
    e) Is successful auto-threading dependent upon wire type?
4) What tapering capabilities are available?
    a) What is the maximum U/V offset in inches?
    b) What is the maximum taper angle in degrees?
    c) Does any part of this feature detract from machine travel?
5) Fine finish options.
    a) What are the finish specifications of my parts?
    b) What are the finish limitations of the machine?
    c) How will this affect machining times?

CAPABILITY

1) Will this machine (properly equipped) do what I need it to do?
    a) For how long?
    b) At what expense?
    c) Is it equipped with the latest technology? (Meaning, "Is it competitive?")

AUTONOMY

1) How long will this machine run unattended?
    a) What is its wire capacity? (Spool weight.)
    b) What is the average rate of wire consumption?
    c) What about wire disposal? (Capacity of disposal system, interval of attention, etc.)
    d) Under my projected operating conditions, what is the efficiency of the threader?
    e) How long will the filters, contacts, resin, etc., last?
    f) What happens if incoming power is lost?

ECONOMY

1) How much does it cost?
    a) How much does it cost to buy?
    b) How much does it cost to run?
    c) How much financial impact will it have on my business?

RELIABILITY
1) Manufacturer's track record.
   a) Total number of machines in the field?
   b) Number of machines of the type being considered?
   c) How long have they been in this business?
   d) References from similar users.

This completes the machine side of the selection process, but there is more to a machine than the iron it's made of. While researching all of the facts and features of the machine-tool itself, you must also seriously consider certain intangibles such as:

SERVICE AND SUPPORT
1) Training (in warranty).
   a) What levels of training are offered?
   b) Number and location of training site(s)?
   c) Number of students allowed?
   d) What kind of follow-up after training can I expect?
   e) Phone support?
2) Training (out of warranty).
   a) Cost?
   b) Availability?
   c) Phone support?
3) Applications support (in warranty).
   a) Field applications? (On-site.)
   b) Number of technicians allotted for this service?
   c) Their average response time?
   e) Phone support?
4) Applications support (out of warranty).
   a) On-site costs?
   b) Availability?
   c) Phone support?
5) Field service support (in warranty).
   a) Number of technicians?
   b) Their location(s)?
   c) Average response time?
6) Field service support (out of warranty).
   a) On-site costs?
   b) Availability?
7) Replacement parts.
   a) Quantity of spares in inventory (dollar amount)?
   b) Location(s) of parts depots or warehouses?

LONGEVITY AND RELIABILITY

1) How long will this machine operate under its expected operation criteria?
   a) Warranty coverage?
   b) Warranty duration?
   c) Are there any exceptions or limitations to the warranty?
   d) Are extended warranties available/recommended/necessary?
2) Are there any records available indicating machine/component reliability? (MTBF charts, part inventory logs, etc.)
   a) What percentage of replacement parts are produced overseas? (Being subject to exchange-rate fluctuations, trade disruptions, duties, tariffs, politics, etc.)
   b) How long will existing U.S. parts inventories last without replacement?

SUPPLIER REPUTATION

1) What is the public's perception of:
   a) The machine?
   b) The dealer?
   c) The builder?
   d) Their individual and collective levels of support?
   e) Their sensitivity to the customer's needs?

REVIEW

That was a lot of work. Congratulations for making it this far! You're not quite finished with this project but you *are* in the home stretch and you're almost ready for a final evaluation before making your selection. We've covered quite a lot of material so let's review it now.

We first had to determine whether this whole exercise was necessary or not. We did this by confirming that we have three things, a:

1) NEED that is qualified by the combination of
2) BUDGET, and sufficient
3) WORK necessary to support it
   After this, we had to examine aspects of the machine's:
4) Size,
5) Speed,
6) Accuracy,
7) Finish,
8) Features,
9) Capability,
10) Autonomy,
11) Economy,
12) Reliability,

13) Service and Support, and

14) Reputation.

As noted earlier, no attempt has been made to list these aspects in any order of importance because each one must be prioritized and weighted by the shop owner, manager, or selection committee. There may be other salient points to consider (go ahead and add them), but the ones we have covered should be enough information to help you make your decision without getting bogged down in the fine print.

SEEING IS BELIEVING

So far, this process should have reduced the field of candidate machines down to a manageable number. From this point on, you must go shopping. By that I mean literally, physically, *go and see the machines*. Now I realize that some people may feel that they "...can't afford to do this." I disagree. What if, because you didn't, you make an expensive mistake by purchasing the wrong machine? Can you afford to do *that?*

Further, the machine builders, importers and distributors go to great expense to present their products at local and national machine shows in order for a potential customer to have every opportunity to see a machine up-close-and-personal. So, should you be unable to journey to one of the larger cities where the builders have technical centers, try to attend a trade-show near you. You will be amazed at how valuable either trip will prove.

While you are at a facility where you can see the machines running, conduct a thorough investigation into the functionality and ease of use of each machine. Anyone connected with the purchase, part planning, and operation of the machine should be present to witness these demonstrations. Many times a potential customer will submit a part to the builder's engineers to be test-cut and returned. Merely sending in a sample part can often be misleading. By not witnessing the test-cut or demo, you have no idea how difficult or easy it was to execute. Attending and watching a demonstration or test-cut will familiarize everyone with what is involved in the programming, set-up and operation of the machine. Simply watching the engineer move and work can give you a very good indication of the degree of ease or difficulty of the machine's daily operation.

If his tasks appeared to be effortless, then you have the confidence that it can be that easy for you also. By being on-site, you will know if he had problems with the machine itself or whether other problems presented themselves.

For example: you may witness a glitchy auto-threader or find that loading one of your existing programs was difficult, or recognize a previously unknown tooling limitation. All of this information and its weight is important when considering what is involved in the daily operation of the machine.

Likewise, the engineer may encounter problems that aren't machine related or indicative of the machine's actual performance. Problems such as material movement due to stress, or a workholding problem. You need to know these things and you can only know them by witnessing the demonstration personally.

By this point, you should have a very good idea as to what your decision will be although there is still one more item to consider. This item may or may not influence your decision but it is important to consider the criticality of the purchase or the:

TIMING

1) How badly do I need this machine?
   a) Urgently. "I am already losing money because I'm not using EDM."
   b) Necessary. "I really need an EDM but we need a new grinder even worse."
   c) Casual. "We've been thinking about EDM to please some of our customers."

Hopefully, your supplier's inventory and training schedules coincides with your sense of urgency.

SUMMARY

Congratulations, you've done it! You should have enough information accrued by now to make an educated evaluation of whether or not you can justify a machine purchase. You may have to examine the financial "cost versus income" portion more closely to better establish the shop rate for this machine, but while you do this you must also be aware of your existing market. (Meaning, just because you have carefully and correctly calculated that the rate for your intended machine should be $75 an hour, this fact won't help you a bit if everyone else in town is doing the same work for $50!)

Should you have any remaining questions regarding the equipment or the details of the transaction, obtain answers from the sales and engineering staffs of your candidate machine builders and distributors. Make them put forth at least the same amount of effort you are making. Require them to answer any and all of your questions — that is their job. Any professional supplier will do their best to accommodate you.

Finally, *make no decision until you have a full understanding of the features and capabilities of both the machine tool and your suppliers.* To do anything else would make this entire survey a waste of time (bad) and possibly jeopardize your entire project (worse).

In closing, I'd offer you luck in your quest for the "perfect" machine, but you must know that in this quest, luck is not a factor. So with this in mind, keep this thought,

Do not allow yourself to be *sold*. Educate yourself so you can *buy*.

SOURCES AND CREDITS

This book was written with the assistance from one or more individuals from the following business firms and educational institutions. To all of these people and sources, my sincere thanks for your help, without which this book's writing would have been much more difficult.

THANKS TO:

Advanced Machining Corporation, NC
Agie USA, Ltd., IL
AmeriWater, OH
Apex Pattern Company, CA
Basic Carbide, PA
Belmont Equipment, MI
Berkenhoff, GmbH, Germany
Carbidie, PA
Charmilles Technologies Corporation, IL
Commonwealth Oil Corporation, Ontario
Composite Concepts Company, OH
CPG Division, MC Machinery Systems, Inc., IL
Dayton Water Systems, OH
Dimensional Control Corporation, CA
E & M Engineering, CA
Ebbco, Inc., MI
EDM Educational Network, CA
EDM Labs, Ltd., CA
EDM Performance Accessories, CA
EDM Sales & Supplies, NJ
EDM Supplies, Inc., CA
EDM Technology Transfer, UT
EDM Video Digest, MI
Electrodes Inc., CT
Eliminator Filter Systems, CA
Eltee Pulsitron, NJ
Erowa, Inc., IL
Everest R & D Corporation, CA
Federal Carbide Company, PA
Feinprüf Corporation, NC
Feinprüf Perthen, GmbH, Germany
Gaiser Tool, Inc., CA
Gisco Equipment, Inc., NY
Global EDM Supplies, OH
Graphel, Inc., OH
Hansvedt EDM, IL
Hausermann Abrading Process Company, IL

Heidenhain Corporation, IL
Ingersoll GmbH, IL
Institute of Advanced Manufacturing Sciences, OH
Intech EDM, IL
Keltool, Inc., MN
Kern Specialty Tools, Inc., CT
LeBlond Makino Machine Tool Company, OH
Leech Carbide, PA
LH Carbide, IN
Master Chemical Corporation, OH
Mecatool, Inc., IL
Methods EDM/Fanuc, Ltd., MA
Mitsubishi EDM, MC Machinery Systems, Inc., IL
Mitsubishi Materials Corporation, CA
Mitsui Machine Technology, Inc., Japax EDM, IL
M.T.I. (Mititoyo, Inc.), IL
MWI Inc., NY
Oberg Industries, PA
North-South Machinery, Inc., CA
Plansee TIZIT, GmbH, Austria
Pacific Controls, Inc., CA
Poco Graphite, Inc., TX
Pystronics, Inc., IL
Rea Engineered Wire Products, IN
Rockwell International, Rocketdyne Division, CA
Sandvik Hard Materials, MI
Sodick, Inc., IL
Sonex Division, Extrudehone Corporation, PA
Spectrum Group International, IL
Swiss Wire EDM, Inc., CA
System 3R USA, Inc., NJ
Task Force Solutions, CA
TechMate International, Inc., IL
Transor Filter USA, IL
T-Star Industrial Products, Inc., IL
UBM Corporation, NJ
University of Nebraska-Lincoln, NE

PAPERS AND PUBLICATIONS

A PRACTICAL GUIDE TO ELECTRODE MATERIAL SELECTION
Poco Graphite, Inc., 1986, 1991

ARTIFICIAL NEURAL NETWORK APPROACH IN MODELING OF EDM
AND WIRE EDM PROCESSES
Indotronix International Corporation, Indurkhya, Rajurkar, and Reddy

CALIFORNIA MANUFACTURERS REGISTER
TMP Publishing Company, 1990

CAPITAL SPENDING SURVEY
Gardner Publications, Inc., 1991, 1993, 1995

CHEMISTRY — AN EXPERIMENTAL SCIENCE
Chemical Education Material Study, Pimentel

EDM TODAY MAGAZINE
EDM Publications

ELECTROLYTIC CORROSION OF WIRE EDMed SURFACES
Precision Engineering Association Journal, Masui, Sone, Takumi,
Seimitsu, and Kogakukai

ENCYCLOPAEDIA BRITANNICA — MACROPAEDIA and MICROPAEDIA
Encyclopaedia Britannica, Inc., Benton

FUZZY THINKING — THE NEW SCIENCE OF FUZZY LOGIC
Hyperion, Kosko

*HIGH QUALITY ELECTRICAL DISCHARGE MACHINING USING AN
ANTI-ELECTROLYSIS POWER SOURCE*
Mitsubishi Electric, Inc., Yamada, Magura, Sato, Yutomi, and Kobayashi

MACHINERY'S HANDBOOK
Industrial Press, Oberg and Jones

METALLURGY AND METALLURGICAL ENGINEERING
McGraw-Hill, Kehl

METALLURGY THEORY AND PRACTICE
American Technical Society, Allen

MONITORING AND CONTROL SYSTEMS FOR DIE-SINKING AND
WIRE EDM PROCESSES
University of Nebraska-Lincoln, Rajurkar and Wang

PLACTICS MOLD ENGINEERING
American Technical Society, DuBois and Pribble

PRINCIPLES OF TUNGSTEN CARBIDE ENGINEERING
Society of Carbide and Tool Engineers, Schneider

RESEARCH AND TECHNOLOGICAL DEVELOPMENTS
IN NON-TRADITIONAL MACHINING
   American Society of Mechanical Engineers, Rajurkar

STANDARD INDUSTRIAL CLASSIFICATION MANUAL
   U.S. Government Printing Office

THE ABC's OF ALUMINUM — FROM RAW MATERIAL
TO APPLICATION
   Reynolds Metals Company, Technical Editorial Service

GLOSSARY

WORDS AND TERMS SPECIFIC TO EDM

Abrading: A process of electrode making. An abrasive-charged epoxy-type model or pattern of the part is made and mounted in a special machine. While reciprocating in a preset circular orbit, it is forced into a graphite blank, effectively "crush-forming" the pattern's shape into the graphite forming the finished electrode. This process is usually reserved for larger, intricate parts such as foundry molds for engine castings and forging dies, etc.

Abrasive Flow Machining: A secondary, postprocessing metal-working operation. Abrasive particles are suspended in semi-liquid polymers of different hardnesses and are forced through or across a material's surface. Used for removing burrs and/or sharp edges from machined parts and removing the recast layer from EDMed surfaces, often in inaccessible areas. (Abbr. "AFM")

ac or Alternating Current: Electric current that flows back and forth in cycles, constantly changing polarity. Opposite of dc or direct current.

Ammeter: A device used for measuring electric current in units of amperes.

Amp/Ampere: The unit of measure for electrical current. Equivalent to the current produced by one volt across one ohm.

Amperage: In EDM, the measurement of average or "working" current during machining. (Abbr. "A")

Angstrofine: In EDM, the term given to a grade of graphite that has a typical particle size of less than 1 micron ($1\mu m$).

Anion: A negatively charged ion. In EDM, the anion resin beads are filled with a strong base and used in conjunction with acid-filled beads to deionize the water used as the dielectric for wire EDM. See: *Cation* and *Mixed-bed Resin*.

Arc: In EDM, a single discharge of current perceived as an electrical spark. Hundreds of thousands of these arcs do the machining in EDM. Commonly confused with "dc arc" which is harmful to both the workpiece and electrode. See: *DC Arc*.

Arc Duration: The length of time EDM current crosses the gap producing work. Usually referred to as "on-time" and is measured in microseconds or μsec.

Automatic Tool Changer: Mechanical device used to augment CNC EDM's ability to run for long periods without operator intervention. Some utilize robotic-type arms to remove/replace the electrodes that run on a tractor-device while others are a shuttle-type system. Often called an "ATC."

Automatic Wire Threader: Mechanical device used to provide autonomy to wire-cut operations. Allows automatic removal and rethreading of a broken wire during machining or automatic wire separation and rethreading when moving from part to part or from detail to detail. Called "AWT" or "AWF" (Automatic Wire Feed).

Automatic Workpiece Changer: Mechanical device used to augment the unattended operation of CNC EDM equipment. Preset parts or workpieces are moved into the worktank for machining and are removed and exchanged for others when completed. Workpieces can be moved by robotic tool or pallet changers or by spindle-type positioning devices. When used in conjunction with an ATC, almost complete machine autonomy can be achieved allowing long-run production or many different parts to be machined unattended. Similar devices can be fitted to wire-cut EDM's.

Average Current: In EDM, the mean (average) current produced during machining, measured at the gap. Most EDM power supplies are rated at average current.

Barrel Effect: The term given to the condition caused by the wire vibrating or resonating within the cut. Where the wire is vibrating the most, it is "thicker" than non- or less-vibrating parts. This usually occurs in the middle of the wire's cutting-length but can resonate and cut "fatter" in more than one place in the Z-height. Secondary discharge can affect part straightness also. Both wire resonance and secondary discharge will affect part straightness. See: *Secondary Discharge*.

Bicycle Effect: The term given to the behavior of the wire during wire-cut operations. In wire EDM, the wire tends to lag behind the wire guides, "cutting the corner," therefore the *bicycle effect* is the difference in the programmed path of the wire guides (the front tire) and the actual path of the wire (the back tire). Also called "corner wash-out." This effect is greater during faster cuts and on taller parts.

Burn, or Burning: Informal expression used to describe the EDM process or machining. For example, "We'll have to *burn* these sharp corners," or "He is *burning* the cavity now."

Capacitor: An electrical component that can store an electrical charge. Used in EDM to increase cutting speed. Also called a *condenser*.

Cation: A positively charged ion. In EDM, the cation resin beads are charged with a strong acid and used in conjunction with base-filled beads to deionize water dielectrics.: See: *Anion* and *Mixed-bed Resin*.

CNC: Computer Numerical Control. In EDM, the control of worktable and/ or electrode motion, automatic tool changers, automatic wire threaders, generator settings, flushing conditions, and other peripheral devices via multiple microprocessors.

Cobalt Depletion: The leaching of the cobalt matrix material that holds the tungsten particles in place. Caused by electrolysis, this typically occurs while wire-cutting carbides using water as a dielectric. Cutting parts while submerged in oil or cut with water along with special anti-electrolysis generator circuitry can alleviate this condition.

Colloidal Suspension: Condition when tiny, insoluble, and nondiffusable particles remain suspended in a liquid. In EDM, these particles are the chips, particulants, and contaminants that are so small they are not captured by the filter media and are too light to "settle out" of the oil or water.

Condenser: Seldom used term for capacitor. See: *Capacitor*.

Conductivity: The comparative rating of a material's ability to carry an electric current. Determined when a material's conductivity is compared to the conductivity of the element silver (Ag) = 100.

Copper: A reddish-brown, malleable, metallic element with a melting temperature of 1082°C that is an excellent conductor of electricity and heat. Shapes of copper and copper alloys and sintered copper forms are used for EDM electrodes. Copper-cores and copper alloys are used in making EDM wire electrodes. Atomic symbol is Cu.

Copper Graphite: A graphite electrode material that has been infiltrated with microfine particles of copper. Used for increased electrical conductivity and electrode strength.

Copper Tungsten: A sintered product composed of tungsten particles infiltrated with microfine copper powder. Used as an EDM electrode material for wear resistance.

Corner Wash-out: Condition occurring during wire-cut operations when the wire lags behind the programmed path and "cuts the corner." This condition is usually encountered during roughing or speed-cuts. See: *Bicycle Effect*.

Corner Wear: Usually the greatest concern of vertical EDM operations and also the most visible aspect of electrode wear. Since corners have the most surface area exposed, they are more susceptible to wear. The sharper the corner angle, the greater the influence of electrode wear.

Crater: The roughly hemispherical shape of the individual cavity left by each spark impact. Consistent of all EDM finishes — rough or fine — whether generated by sinker or wire.

Cubic inches/hour: One measurement of the rate of EDM metal removal. (Abbr. in.^{3}/hr. or cu. in./hr.) See: *Metal Removal Rate*.

dc Arc: In EDM, the continuous, uncontrolled flow of current between the electrode and workpiece caused by poor flushing, gap contamination, short ionization-times, etc. Often even the smallest of dc arcs can severely damage the workpiece and electrode. Not to be confused with "arc," which is actually a good, controlled spark. See: *Arc*.

dc or Direct Current: Electric current that always flows in a single direction, maintaining a constant polarity.

Deionization: In EDM, the process of removal of conductive ions from the water dielectric by passing it through a container of resin beads infiltrated with cation and anion exchangers. See: *Cation* and *Anion*.

Deionize: To make a substance nonconductive, in this case, the EDM dielectric.

Deionized Water: Used as the dielectric in wire EDM. Water that has had all salts and metallic ions chemically removed by the machine's resin system. See: *Resin*.

Die Sinker: See: *Sinker*.

Dielectric: A material that does not conduct electricity but can sustain an electric field. In EDM it is the dielectric fluid that acts as an insulator until a certain distance (spark gap) and voltage are achieved, then it ionizes and supports the electric field through which machining current passes. Traditional dielectrics used in EDM are special "oils" for vertical machines and water for wire machines although both have been used in each machine type.

Dielectric Strength: The ability of a dielectric to resist voltage break down and maintain its insulating qualities against premature electrical discharge. Its rating would be the minimum voltage required to produce an electrical spark through a dielectric medium. A high dielectric strength is a desired property of an EDM oil.

Discharge: In EDM, the controlled flow of current across the gap producing a spark.

Discharge Column: The ionized, electromagnetic flux or field that permits the EDM spark and the flow of electrons during EDM machining.

Discharge Dressing: Process used to make or dress metallic electrodes in the EDM machine. This procedure is possible in almost all properly equipped EDM machines, but is easiest to implement in CNC EDM, increasing autonomy, machine-tool utilization, and cost savings through reduced direct-labor.

Dither: A device on some EDM machines that provides a high-frequency vibration to the machine quill and electrode to aid in flushing and, hence, stability.

Dress/Redress: To physically remove the worn areas from an EDM electrode. To "sharpen" an electrode. The dressing and redressing of electrodes can be performed by conventional machining, discharge dressing and certain abrasive methods. See: *Abrading* and *Discharge Dressing*.

Duty Cycle: This is the measure of efficiency of an EDM frequency. It is the percentage of on-time relative to the sum of on-time and off-time, calculated as follows: *On-time + off-time ÷ on-time x 100 = Duty Cycle*.

Edge-start: In wire-cut EDM, the machining of any part when entering from an outside edge rather than a hole or opening. A punch for a die would be an example of a part starting with an edge-start.

Efficiency Meter: A gauge or meter on some power supplies that provides the operator with a visual reference measuring the duty cycle or the efficiency of EDMing. See: *Duty Cycle*.

Electrical Discharge Grinding: A process using EDM theory to machine electrically conductive materials. These machines resemble a surface grinder or horizontal mill more than an EDM machine. A rotating graphite wheel is used as an electrode as the workpiece is slowly fed into it in a manner similar to creep-feed grinding. (Abbr. "EDG") See: *Electrical Discharge Machining*.

Electrical Discharge Machining: The carefully controlled process of metal removal using a series of electrical sparks enveloped within a dielectric fluid to machine electrically conductive workpieces. Material shapes are produced in workpieces by either a shaped electrode or a "travelling" wire. Also called "*spark erosion*." (Abbr. "EDM")

Electrode: The tool or "cutter" that allows the transfer of electrical energy to EDM a workpiece. It must be electrically conductive and can be a shaped electrode as used in a vertical EDM, or a wire used in wire-cut EDM.

Electrolysis: Electrolysis occurs in the presence of dissimilar metals, water, and electric current and results in the decomposition of a material by the action of an electrical field. In EDM, electrolysis contributes to rusting of ferrous materials, oxidation of nonferrous materials, bluing of titanium, and the pitting and cobalt depletion of carbides.

End Wear: Electrode wear that results in the reduction of the length of an electrode.

Farad/microfarad (μfarad): The measurement or unit of electrical capacitance. Equal to the amount that permits the storing of one coulomb of charge for each volt of applied potential.

Filtration: The important process of removing chips, debris, and contaminants from the dielectric fluid by passing it through a porous filter media. This media can be paper, cloth, charcoal, sand, diatomaceous earth, etc. There are also centrifuge and electrostatic filtration systems in use. Clean dielectric supports faster, safer cutting and better finishes and increased resin life.

Finish: The EDM finish is a random array of millions of tiny craters having no linear or circular orientation. The finish callout on the blueprint is referring to the standard or scale of finish desired: R_{max}, $\mu m R_{max}$, RMS, R_a, VDI, etc.

Finish Cut: In EDM, the term used for a final pass of the wire or electrode to size and finish an EDMed detail. This is typically done with reduced current and high frequencies.

Flashpoint: In EDM, the temperature at which an EDM oil, at a specific vapor/air ratio, will support momentary combustion. Simplified — the higher the flashpoint, the safer the EDM oil.

Flexural Strength: The ability of a solid material to withstand a flexural or transverse load. A concern when making or using very thin or delicate electrodes.

Fluid Level Switch: The safety float switch on a vertical EDM that switches the current off in the event the dielectric drops below a specific level. This prevents exposing the spark gap to oxygen while under power and reduces the probability of fire.: On wire-cut EDM, this switches the fast-fill pumps off, preventing the upper head from being submerged.

Flush Cup/Nozzle: On a wire-cut EDM, the removable/replaceable nozzle through which the wire and water pass. So named because they direct the flow of water used in flushing. These are usually made of a nonconductive plastic or ceramic.

Flush Pot: A tooling/flushing device used in vertical EDM. A hollow worktable to which workpieces can be mounted over an opening(s) in the table, providing pressure or vacuum flushing. It can also allow the electrode to pass completely through the part if necessary. Also called a *plenum*.

Flushablilty: Term given to evaluate EDM wire-types for flushing capabilities. Some wire alloys are better than others. For example, zinc-coated wires have good *flushability* while molybdenum wires do not.

Flushing: The process of physically forcing or exchanging contaminated dielectric from the spark gap with fresh clean dielectric. See: *Sealed Flushing*.

Flushing hole: Hole through which dielectric oil is forced. This hole can be in either the electrode or the workpiece.

Frequency: The number of on/off cycles in one second. Measured in units called hertz, equal to one cycle per second. Calculated as follows: *1000 ÷ (on-time + off-time) = frequency in KHz.*

Fuzzy Logic: Advanced computer technology used in many high-end, decision-making computers and micro-processors. Fuzzy logic now augments some EDM controls allowing it to "think" its way through difficult or changing machining conditions.

Galvanic: The flow of electrons in direct current. Used to electrically deposit zinc coatings to EDM wire electrodes. As in *"galvanize."*

Galvano Method: Electrode-making process where heavy deposits (up to 0.200" thick) are electrolytically deposited or plated onto a plaster or epoxy pattern. Used to make large electrodes when electrode weight would be too heavy to allow good servo response if formed entirely from solid copper. Also used when electrode material cost is a factor because of its large size.

Gap: See: *Spark Gap*.

Gap Voltage: Measurement of electrical voltage across the gap between the electrode and workpiece. When the electrode or wire is too far away to allow machining or current, this is referred to as "open-gap voltage" and is typically very high, up to 300 V. During EDM machining, it is called "working-gap voltage" and is much lower, as low as 35 V.

Generator: EDM power supply.

Graphite: A carbon-based electrode material. Classified as a *metalloid* because it exhibits certain properties and characteristics of a metal yet it is not a actually a metal. EDM graphites are made from various grain sizes of carbon compounds (amorphous carbon) and mixed with coal tar and other binder materials. It is then baked at extremely high temperatures until all volatiles and gases have been driven off and all remaining components have been "graphitized." Graphite as an electrode material provides EDM users with ease of fabrication, good metal removal rates (MRR), and high resistance to wear.

Harmonics: The vibration or resonance of the wire that occurs during wire-cut operations. Contributes to the condition known as the "barrel-effect."

Head: The servo-controlled vertical or Z-axis of an EDM machine. This part is also called the *ram*, hence the name, *ram EDM*. The upper and lower control-arms of wire EDM's are often referred to as the upper or lower *head*.

Heat Affected Zone: The area immediately below the recast layer that has been influenced by heat from the EDM process. Its depth depends greatly upon the workpiece material and EDM conditions. (Abbr. "HAZ")

Hertz: The international measure of unit of frequency. Equal to one complete electrical wave cycle in one second. (Abbr. "Hz")

High-Pressure Flushing: In wire EDM, forcing the water dielectric into the gap at pressures of up to 300 psi. High-pressure flushing contributes significantly to high-speed cutting.

Hydrostatic Press: A device used for compacting carbide powders. Metal powders are contained within a rubber bladder and compressed hydraulically with pressures from 25,000 to 30,000 psi. In carbide making, *hydrostatic* generally refers to the use of a liquid medium (oil) at room temperature. See: *Isostatic Press*.

Infiltrated Graphites: Specialty graphites that have been blended with varying percentages of micro-fine copper powder before sintering for increased conductivity and electrode strength. See: *Copper Graphite*.

Initiation Voltage: The amount of voltage at a specific gap that is required to overcome the resistivity of the dielectric to initiate the EDM spark.

Insulator: Any substance that prevents the flow of electric current. In EDM, this can be ceramics, "green glass," delrin, synthetic rubies, etc. These can be actual machine components, or used in tooling and fixtures.

Ion: The electrical condition of a previously neutral atom or molecule that has assumed a definite electrical charge (positive or negative) after gaining or losing one or more electrons. The EDM spark travels within an *ionized* channel or conductive path.

Ionization: In EDM, when a path of molecules within the dielectric become polarized or aligned, allowing resistivity to drop and current to flow between the electrode and workpiece.

Ionization-time: The amount of time it takes the presence of voltage to sufficiently ionize the dielectric to support the flow of current. The amount of time required depends a great deal upon the fluid's dielectric strength and flushing conditions.

Ionization Voltage: The voltage at which current begins to flow across the gap. Typically higher than working voltage.

Isostatic Press: A device used for compacting carbide powders. In carbide-making, *isostatic* is term used for describing the device using gas at elevated temperatures. Hot, isostatic pressing is used to remove porosity after sintering. See: *Hydrostatic Press.*

Isotropic: In EDM, the description of the "grain structure" of graphite electrode material. Isotropic graphites will have a uniform, omnidirectional structure without a grain, allowing electrodes to be made without regard to its original orientation within the graphite blank. Opposite of *Anisotropic.*

Low Wear: Term given to generator settings that provides good machining speed with minimal electrode wear. See: *No Wear.*

Metal Removal Rate: The measurement of material removal by EDM, usually expressed in cubic inches per hour or in.^{3}/hr. (Abbr. "MRR")

Micro-ohm: One-millionth of an ohm. Written as $\mu\Omega$. See: *Ohm.*

Microinch: A unit of length measuring one-millionth of an inch. Written as μin.

Micrometer: A unit of length measuring one-millionth of a meter. Written as μm.

Micron: One micrometer. Equal to 0.000040" or 40-millionths of an inch. Written as μm. See: *micrometer.*

Microsecond: One-millionth of a second. Written as μsec.

Mixed-bed Resin: In EDM, the mixture of small (< 1 mm) polyester beads that have been infiltrated with a strong acid or base and used to deionize EDM water. The ratio of base (anion) to acid (cation) is 60:40 respectfully, hence the name *mixed-bed* resin. See: *Anion* and *Cation.*

Multilead: A type of power delivery used when EDMing with multiple electrodes simultaneously. By connecting individual power leads to each electrode, faster machining speeds can be obtained.

No Wear: Term given to EDM machining when protecting the electrode against wear by using elevated on-times. Modern generators can deliver wear characteristics of <0.1% by volume in metallic and graphite electrodes.

Off-time: The "rest" part of the spark cycle. Required to allow reionization of the dielectric and aid in flushing. The duration of this increment significantly affects the overall speed and stability of the machining operation.

Ohm: The unit of measure of electrical resistance. One ohm will allow one volt to maintain a current of one ampere. Symbol: Ω.

Omnidirectional: Referring to a finish having no grain or lay as in the EDM finish. The EDM finish is a random array of millions of tiny craters having no linear or circular orientation.

On-time: The "work" part of the spark cycle. Current flows and work is done only during the on-time. The duration of this increment will significantly affect the overall machining speed, workpiece finish, and electrode wear.

Open Gap Voltage: Maximum voltage potential before ignition and flow of current. Seen on the voltmeter when the start button is pushed and electrode or wire is approaching the workpiece.

Orbit: A preselected or programmed pattern or shape of movement to be used to impart motion to the electrode or workpiece.

Orbiter: A mechanical device that can be fitted to the head of a manual EDM to provide electrode motion to aid machining.

Orbiting: The use of circular or other symmetrical movement of the electrode or worktable used to size and finish a workpiece detail. This orbit is usually a "canned cycle" around the centerline of the workpiece detail rather than a hand-programmed movement. *Vectoring* out from center is another capability.

Oscilloscope: Device used to "see," measure and evaluate the current, spark frequency, and wave form during machining.

Overcut: The allowance that must be made for the size of the spark in addition to the size of the electrode or wire. Also called *overburn, spark gap,* or *kerf.*

Particle Size: In EDM, usually associated with the particle or grain size of graphite electrodes. The smaller the grain, the higher the quality.

Peak Current: The highest current an EDM generator can briefly produce. Peak current is the measure of intermittent "spiking" of amperage. By mathematical formula, a 35 amp (average) EDM generator can briefly peak at 960 amps. Not to be confused with *average current.*

Plenum: See: *Flush Pot.*

Polarity: The direction of ion alignment and current flow. In EDM, this term usually refers to the electrode's polarity — positive or negative.

Power Supply: Component of the EDM system that contains the generator, transistors, capacitors, and other electrical devices required to produce the EDM spark. Also known as a *generator.*

Pressure Flushing: Referring to vertical EDM, a common and usually the most efficient method of "sealed flushing" used. Accomplished by forcing the dielectric oil, under low pressure, through the spark gap via holes in the electrode or the workpiece. Also called "injection" flushing. See: *Sealed Flushing.*

Pulse: In EDM, a brief surge of electrical current. The current used in EDMing must be *pulsed* (turned on and off) to provide time to allow the dielectric to reionize.

Pulsed Flushing: The timing of pressure or injection flushing with the peck-cycle retraction of the electrode from the cavity.

Ram: The vertical positioning component of a vertical EDM. Also called the "head."

RC Circuit: Older, capacitor-based generator circuitry. RC is short for Resistor Capacitance, but has also been called a "relaxation circuit" because when compared to transistorized generators, capacitance discharge is slow, seemingly "relaxing" more than immediately switching on and off. This results in an angular, "saw-tooth" shaped wave form instead of the precise square-wave produced by transistors.

Recast/Recast Layer: The name of the condition of the EDMed surface after having been melted then resolidified or "re-cast" back onto the parent material during the off-time part of the EDM cycle. The recast layer is metallurgically very different than the parent material, and is also influenced by the type of electrode and dielectric used.

Resin: Term given to the polyester beads containing the strong chemicals (acids and bases) that condition the water dielectric used in wire EDM. See: *Mixed-bed Resin.*

Resistivity: In wire EDM, the measurement of the resistance of the water dielectric. Opposite of conductivity.

Resolution: In EDM, the measurement of axes positioning via servo-motor encoders, resolvers, or linear-scale feedback. The finer the resolution of the "reader system," the smaller the increment of movement.

Roughing: Term given to rapid EDMing of a part or cavity. Opposite of finishing.

Sealed Flushing: So-called when the advancing face of an electrode allows EDM oil to pass only through the spark gap. Flushing is most efficient when it is "sealed" in this manner. Sealed-flushing applies to wire EDM when the flush cups or nozzles can be positioned within 0.002" - 0.005" of the workpiece.

Secondary Discharge: Situation when conductive or semiconductive particles suspended within the dielectric touch the sides of the electrode or workpiece and effectively reduce the spark gap causing random discharges upon the previously finished workpiece surface. This can cause pitting, irregular finishes, and tapered cavities. Secondary discharge in vertical EDM can be prevented by using vacuum flushing instead of pressure flushing. This situation also contributes to the "barrel-effect" in wire EDM operations. See: *Vacuum Flushing* and *Barrel Effect.*

Servo-system: The driving mechanism of any controlled axis (axes). This can be either hydraulic or electrical.

Side Wear: The measurable wear or taper on the sides of an electrode.

Silicon Powder: In vertical EDM, certain silicon powders can be added to the dielectric oil to diffuse a single EDM spark into several smaller ones for quicker, fine finishing of EDMed cavities. Some wire machines also have this capability.

Silver Tungsten: A special sintered electrode material similar to copper tungsten, only silver particles are used instead of copper. Used when higher conductivity is desired.

Sinker/Die Sinker: A common name for vertical EDM. Taken from the antiquated cavity-making process called *hobbing*. Not to be confused with *gear-hobbing*, hobbing by this definition is the process of forcing a prehardened tool steel shape or hob into the workpiece material under great pressure, effectively cold-forming a cavity for a die or mold. This process was called *sinking a die* or *die sinking*, so the term "sinker" for EDM evolved naturally because an *electrode* was sunk into the workpiece instead of a hardened hob.

Skim-cut/Skimming: Term given to low-power cuts or passes in wire-cut operations used for finishing and sizing a part or detail. Also called *trim-cut/trimming*.

Slug: The scrap material that drops out of the workpiece after wire-cutting or in sinker operations, trepanning. Unless the detail is quite small, the ability to "drop a slug" is far more efficient than "pocketing" or EDMing the entire volume of a shape.

Slurry: An abrasive liquid or semi-liquid mixture of water or oil and abrasive compounds used in ultrasonically dressing/redressing graphite electrodes.

Sonotrode: In EDM use, it is the rapidly vibrating (up to 20,000 Hz) component of an electrode-making ultrasonic machine to which the master pattern is affixed. Also called the *horn*.

Spark: The electrical discharge between any charged conductor and a ground. See: *Arc*.

Spark Erosion: Another name for the EDM process. See: *EDM*.

Spark Gap: The distance between the electrode or wire and the workpiece during machining. All information and electronic feedback to the control regarding servo advance and retract is obtained in the spark gap. This information includes parameters such as machining current, servo voltage, gap contamination flushing conditions, etc.

Split Electrode: Multiple-electrode configuration. Can be used with either single or multilead power supplies. See: *Multilead*.

Split-lead: See: *Multilead*.

Square-wave: Term given to the spark's digital wave form from a transistor power supply. Today's technology allows spark wave forms to be modified and altered, seldom remaining "square." Opposite of sinusoidal.

Start Hole: A predrilled opening in a workpiece providing a place for threading the EDM wire.

Stress-cuts: Special wire-cuts placed strategically around the final part configuration before it is machined to free it from possible influence or movement caused by any stresses resident within the surrounding material.

Sublime/Sublimate: The process where a material at elevated temperatures, passes directly from a solid to a gas, bypassing the liquid state entirely. Graphite electrodes *sublimate*.

Submerged Machining: Type of machining used in wire-cut operations when the part is completely submerged by the dielectric to aid in flushing. Good part candidates for submerged machining are parts with steep tapers, parts with varying thickness, and parts with intersecting openings and interrupted cuts. Edge-starts are made easier when the part is submerged.

Suction Flushing: See: *Vacuum Flushing*.

Surface Finish: Referring to the roughness or smoothness of a given surface.

Surface Integrity: The metallurgical structural quality of the EDMed surface and subsurfaces.

Swarf: Term given to the carbonized particles generated by EDMing.

Tap Buster: Simple machine using the EDM process that is used only for EDMing holes or to remove broken cutting tools from workpieces. So-named after their first uses.

Tellurium Copper: Copper electrode material that has a small amount of the element tellurium added for machinability.

Through-hole Flushing: Flushing procedure using electrodes or workpieces that forces the EDM oil directly through the spark gap instead of by external means.

Transducer: A device that changes electromagnetic forces into mechanical force.

Transistor: An electrical component used as a high-speed switching device. Most modern power supplies are *transistorized*.

Trepan: In EDM, the machining process where a hollow electrode is used to remove large cores or slugs from workpieces without reducing the entire cylinder volume to small chips.

Tungsten: An element with very high hardness and melting temperature used within different matrixes as cutting tools, wear parts, electrical contacts, etc. As an electrode material it is used in the form of solid wires or rods for small-hole drilling or is alloyed with copper or silver to provide increased wear resistance of electrodes. Its very high melting point of 3370°C, provides its good wear resistance as an electrode. (Atomic symbol is W for Wolfram.)

Vacuum Flushing: The process of using vacuum to draw clean, cool dielectric oil through the spark gap. Opposite of pressure flushing. Also called suction flushing. See: *Secondary Discharge*.

Viscosity: The measure of a liquid's resistance to flow. The higher the viscosity, the thicker the liquid. In EDM oils, lower viscosity is usually desired, especially when fine finishes and sharp detail are required.

Volt: The difference in electric potential between two points in an electric field that requires one joule of work to move a positive charge of one coulomb from the lower potential to the higher one.

Voltage: The term given to electromotive force or the difference in electric potential and expressed in volts.

Voltmeter: A device used for measuring voltage.

Volumetric Wear: The total amount of electrode wear combining end, corner, and side wear, expressed in cubic inches.

Wave Form: In EDM, the shape or "signature" of the EDM spark or discharge. Waveform shapes and intervals can be controlled by generator settings and can be seen on an oscilloscope.

Wear: The thermal consumption of the EDM electrode during machining. There are four types of wear: end, corner, side, and volumetric.

Wear Ratio: The comparison or ratio of electrode wear to the amount of material removed.

White Layer: The term given for the thermally changed EDMed surface. This condition can vary with different workpiece materials, electrode materials, and the type of dielectric used. The workpiece surface hardness can be affected resulting in hardening or annealing of the EDMed surfaces. See: *Recast/Recast Layer*.

Wire Alignment Block: Precision block used to set the EDM wire perpendicular to the worktable. Manual alignment blocks are rectangles of hardened and ground stainless steel. An automatic wire alignment block is a special electrical unit that is connected to the machine's control. The control can then automatically adjust the position of the wire and set zeros in the *U/V* axes readouts.

Wire EDM: Utilizing the same basic EDM theory as a vertical EDM, it is a specialized machine that uses electric current passing through a wire electrode to cut precision shapes in conductive materials.

Wire Guide: Precision diamond or sapphire bearings that control and guide the wire according to the CNC program. Some guide types are round "donuts" and others are split-vee types.

Working Voltage: The measure of voltage during the cutting process. Opposite of *open gap voltage*.

X-Axis: When standing in front of a machine tool, it is left and right movement in the horizontal plane.

Y-Axis: When standing in front of a machine tool, it is forward and back movement in the horizontal plane.

Z-Axis: When standing in front of a machine tool, it is an up or down movement in the vertical plane.

APPENDIX A

SUMMARY PROCESS ANALYSIS AND COST COMPARISON STUDIES

SUMMARY PROCESS ANALYSIS AND COST COMPARISON

3-PIECE ACTUATOR LINK
A-SERIES AIRBUS

MATERIAL:
Inconel 718

OBSERVATION:
Present manufacturing methods using conventional milling operations have proven inefficient and cost-prohibitive.

OBJECTIVES:
1) **Reduce setup time:**
   Multiple, labor-intensive set-ups in use require expensive tooling.
2) **Reduce unacceptably long machining times:**
   Currently 104 hours per set.
3) **Reduce unacceptably high scrap rates:**
   Caused by a combination of actual machining and set-up errors, inaccuracies due to cutter and part deflection, damaged parts caused by cutter breakage, and unpredictable material movement during machining.
4) **Reduce or eliminate secondary operations:**
   a) Deburring
   b) Stress-relieving
   c) Heat-treating
   d) Straightening
   e) Jig-grinding
5) **Reduce extensive cobalt cutter inventory:**
   Part requires carbide inserted fly-cutters, numerous expensive "corn-cob" and conventional cobalt end-mills, and carbide-tipped spade drills.
6) **Reduce operator involvement:**
   Multiple set-ups are labor-intensive and time-consuming, and increase the opportunities for errors.

287

RECOMMENDATIONS:

All objectives can be met by producing these parts using a high-speed wire-cut EDM.

CORRECTIVE ACTION:

1) **Reduce set-up time:**

   Is significantly reduced to one simple set-up allowing 90° indexing to machine second axis, thereby eliminating all of the expensive milling tooling.

2) **Reduce machining times:**

   A wire-cut EDM equipped with an automatic wire threader affords unattended, single-pass machining with common parting lines to reduce machining times to less than 19 hours, complete.

3) **Reduce scrap:**

   The combination of reliable and repeatable CNC provides predictable, accurate, and repeatable results and virtually eliminates any possibility of scrapping expensive parts.

4) **Reduce or eliminate secondary operations:**

   a) **Deburring:** Eliminated

   Wire EDM produces parts with clean, sharp edges requiring no post-processing operations to remove burrs.

   b) **Stress-relieving:** Eliminated

   The EDM process imparts no work-hardening, cutter-stress, or related heat damage to influence the desired condition of the material.

   c) **Heat-treating:** Eliminated

   Parts to be EDMed can be machined "in the hard" or in their final state, as per print.

   d) **Straightening:** Eliminated

   There are no cutting pressures to deflect or deform thin or weak sections. In addition, all three pieces are machined as a set, from the same grain-oriented blank, further reducing any unpredictable material movement.

   e) **Jig-grinding:** Eliminated

   All through-holes are done while all three pieces are still captive and stacked together, affording the straightness and cylindricity of line-boring along with providing "split-tenth" accuracy and appropriate finish.

5) **Reduce cutter costs:**

   Almost all cobalt cutters previously used in conventional machining have been eliminated. Even the initial fly-cut squaring operation has been eliminated by using raw bar stock. Only wire-start hole drilling operations and a single milling operation to machine the "eyebrow" counterbore on two parts of the three part set remain.

6) **Reduce operator involvement:**

Utilizing a wire-cut EDM equipped with an automatic wire threader practically eliminates operator involvement and affords extended hours of unattended machining.

RESULTS:

Changing the manufacturing process from conventional milling to wire-cut EDM has:

1) Reduced multiple set-ups and related tooling to a single setup allowing one 90° indexing.
2) Reduced machining times from 104 hours to 19 hours per set.
3) Virtual elimination of scrap and rework.
4) All post-processing operations eliminated.
5) An $18,000 per month savings through reduction of cobalt cutting tool inventory.
6) Significantly reduced direct-labor costs through sustained, unattended operation.

BONUS:

7) Waste material is in slug form, clean and dry, instead of oily chips subject to EPA scrutiny. These slugs are easily handled and are recyclable at a high dollar scrap rate.

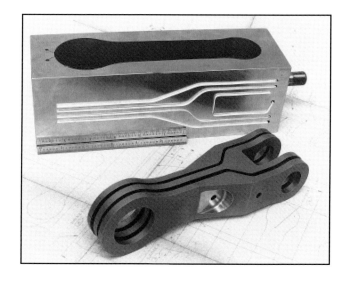

SUMMARY PROCESS ANALYSIS AND COST COMPARISON

LAPPING TOOL USED IN PRODUCING DENTAL DEVICE

MATERIAL:
ASM 17-4PH Stainless Steel

OBSERVATION:

Present manufacturing methods require expensive 5-axes machining and an extensive inventory of cutting tools. The present methods are operator intense, require multiple secondary operations and the total cycle time required to produce parts is unacceptable.

OBJECTIVES:

1) **Reduce cycle time for a finished part:**
   Machining time for one part is 1 hour and 16 minutes and still requires an additional 3 hours of manual deburring and polishing.

2) **Reduce high cost of machining 5-axes contours:**
   Part has compound angles and spherical radii and is pick-feed machined on a high-speed spindle Bostomatic CNC milling machine, using very small cutters and a very fine feed. Machining time and surface finish are unacceptable.

3) **Reduce inventory of expensive cutting tools:**
   Part machining requires numerous small diameter, high speed radiused end mills. These cutters are very small, subject to breakage and are difficult to resharpen.

4) **Reduce operator involvement:**
   The rapid consumption of perishable cutting tools requires the almost constant presence of an operator to monitor progress and change cutters.

5) **Reduce or eliminate secondary operations:**
   a) Deburring
   b) Polishing
   c) Post-process heat-treating

RECOMMENDATIONS:

This part can be economically produced using a vertical EDM with an integral C-axis to roll-form the impressions using electrodes and tooling produced on a wire-cut EDM.

CORRECTIVE ACTION:

1) **Reduce part cycle time:**
   Roll-forming all details using vertical EDM is far more efficient than single-point machining. All surfaces are machine polished and finished in the machine in under seven minutes.

2) **Reduce high cost of producing part:**
   A wire-cut EDM machine enables the unattended production of copper electrodes with close-tolerances and predictable repeatability. These electrodes are then mounted in an electrode holder (also produced in the wire machine) and used to roll-form the complex impressions using the C-axis equipped vertical EDM machine.

3) **Reduce cutter costs:**
   The need for numerous expensive, small diameter end mills has been eliminated entirely.

4) **Reduce operator involvement:**
   Utilizing a wire-cut EDM and linking machining programs and "stringing" parts allows electrode production to run unattended. By stacking multiple parts in the vertical EDM, numerous finished parts can be produced with little or no operator involvement.

5) **Reduce or eliminate secondary operations:**
   a) **Deburring:** Eliminated
       Vertical EDM produces parts with clean, sharp edges requiring no post-processing operations to remove burrs.
   b) **Polishing:** Eliminated
       The EDM finish on the finished part is smooth and free of all cutter and machining marks, eliminating secondary polishing operations. Multiaxes machining while roll-forming this configuration provides the finish required by vendor.
   c) **Post-process heat-treating:** Eliminated
       Parts can be EDMed "in the hard" requiring no outside post-processing.

RESULTS:

Changing the manufacturing process from conventional milling to wire-cut and vertical EDM has effectively:

1) Reduced machining time for a single part from 1 hour 16 minutes to just under seven minutes.
2) EDM finish meets blueprint requirements, "as-is." This eliminates all deburring and polishing operations, reducing cycle time an additional 3 hours.
3) Reduced high cost of production. The total cost for the electrode holder, copper electrode material, finished electrode inserts and programming time is $850.00, completed in only 16.5 hours.
4) The elimination of all expensive cutting tools through the use of easily produced copper electrode inserts. Electrode wear was held to 0.1% and, in production using "stacked" parts, produces 27 part details per electrode.
5) Significantly reduced direct-labor costs through sustained, unattended operation.

(*Summary Process Analysis and Cost Comparison* courtesy of North-South Machinery, Inc.)

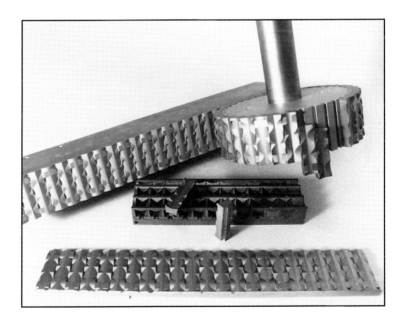

SUMMARY PROCESS ANALYSIS AND COST COMPARISON

HYDRAULIC PISTON ASSEMBLY
AIRCRAFT NOSEWHEEL STEERING UNIT

MATERIAL:
   AISI 440 Stainless Steel

OBSERVATION:

   Present manufacturing method using conventional milling operations is much too slow and requires excessive amounts of costly end-milling. Present method creates multiple secondary operations and is not cost effective. Wire EDM would normally be specified except for the 120° spacing of the slots prohibit "through-hole" machining usually associated with wire-cut machining.

OBJECTIVES:

1) **Reduce machining time:**
   Alloy is difficult to machine, in addition to slot detail being narrow and deep — warranting slow feed rates.
2) **Reduce high scrap rates:**
   Caused by a combination of; actual setup and machining errors, inaccuracies due to cutter-whip or cutter deflection, and gouging damage caused by broken end-mills.
3) **Reduce or eliminate secondary operations:**
   a) Deburring
      Large, tough burrs are left on outside and inside diameter of part, with limited access to the inside diameter.
   b) Polishing
      Milling leaves unacceptable machining marks that must be removed either by manual polishing or AFM (Abrasive Flow Machining).
4) **Reduce costly special cutter inventory:**
   Part requires numerous extra length cobalt and carbide end-mills.

5) Reduce operator involvement:

Due to the difficult milling application, operator attendance is required to monitor conditions and frequently change cutting tools.

RECOMMENDATIONS:

These objectives can be met by producing these parts using a high-speed wire-cut EDM machine fitted with an aftermarket C-frame device that is commercially available and relatively inexpensive.

CORRECTIVE ACTION:

1) Reduce machining times:

A wire-cut EDM machine equipped with a C-frame attachment allows wire-cutting of this part because of the ability to present the wire to only one side of the diameter.

2) Reduce scrap:

Scrap reduction is attained via repeatable CNC programming and the total elimination of cutter-whip, deflection, and breakage inherent to conventional machining.

3) Reduce or eliminate secondary operations:

    a) Deburring: Eliminated

    Wire-cut EDM produces parts with clean, sharp edges requiring no post-processing operations to remove burrs.

    b) Polishing: Eliminated

    The finish left by wire EDM is smooth and leaves no machining marks as with all other "conventional" methods.

4) Reduce cutter costs:

All cobalt and carbide end-mills previously used to conventionally machine this part have been eliminated. Three, non-precision wire-start holes are required 120° apart and they are the only other operation required other than wire-cutting this detail.

5) Reduce operator involvement:

A wire-cut EDM equipped with an after-market C-frame device will substantially reduce operator involvement while providing extended hours of unattended machining. While an automatic wire-threading device would normally allow many applications to run totally unattended, this is not a viable consideration due to the design and configuration of the C-frame. Operator involvement is limited to manually threading of the wire three times while the slot itself is cut completely unattended.

RESULTS:

Changing the manufacturing process from conventional milling to wire-cut EDM has:

1) Reduced machining times by almost 50%.
2) Virtually eliminated scrap and rework.
3) Post-processing operations of deburring and polishing have been eliminated.
4) The elimination of all conventional cutting tools except inexpensive twist drills.
5) Significantly reduced direct-labor costs through sustained, unattended operation.

(*Summary Process Analysis and Cost Comparison* courtesy of North-South Machinery, Inc.)

APPENDIX B

EDM PURCHASE PLANS BY INDUSTRY AND COMPANY SIZE

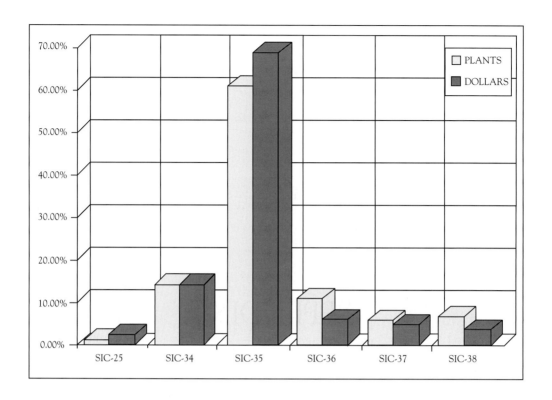

PLANTS PLANNING TO PURCHASE RAM EDM BY 2-DIGIT SIC'S

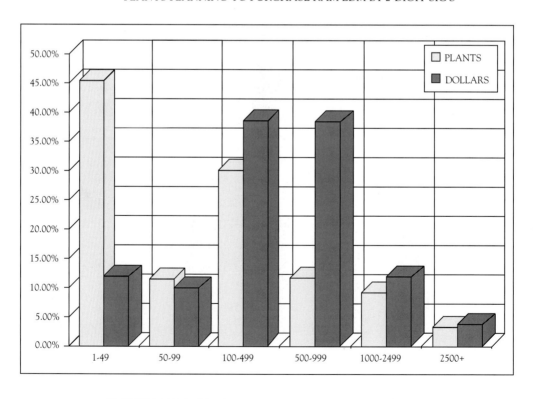

PLANTS PLANNING TO PURCHASE RAM EDM BY SIZE (EMPLOYEES)

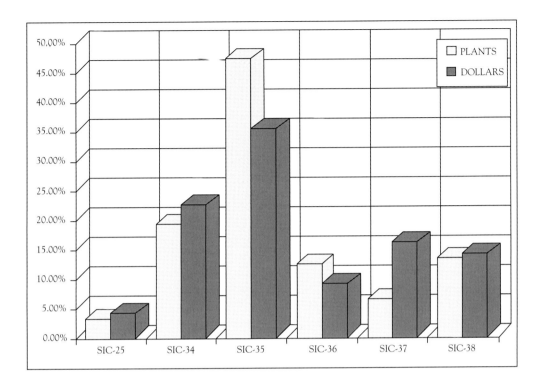

PLANTS PLANNING TO PURCHASE WIRE EDM BY 2-DIGIT SIC'S

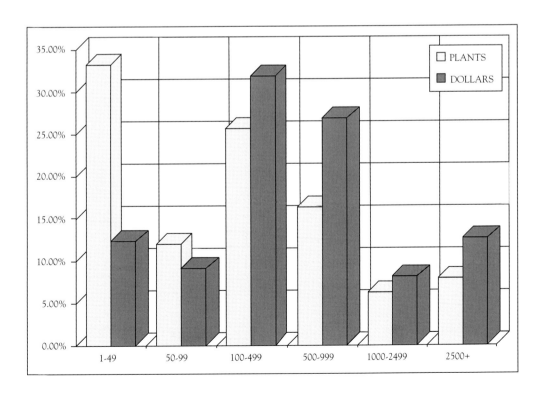

PLANTS PLANNING TO PURCHASE WIRE EDM BY SIZE (EMPLOYEES)

INDEX